Alexander Bilmes

Resolving locations of defects in superconducting transmon qubits

Experimental Condensed Matter Physics
Band 27

Eine Übersicht aller bisher in dieser Schriftenreihe erschienenen Bände finden Sie am Ende des Buchs.

Resolving locations of defects in superconducting transmon qubits

by

Alexander Bilmes

KIT Scientific Publishing

Karlsruher Institut für Technologie
Physikalisches Institut

Resolving locations of defects in superconducting transmon qubits

Zur Erlangung des akademischen Grades eines Doktors der Naturwissenschaften von der KIT-Fakultät für Physik des Karlsruher Instituts für Technologie (KIT) genehmigte Dissertation

von Dipl.-Phys. Alexander Bilmes

Tag der mündlichen Prüfung: 08. Februar 2019
Referent: Prof. Dr. Alexey V. Ustinov
Korreferent: Prof. Dr. Alexander Shnirman

Impressum

KIT Scientific Publishing

Karlsruher Institut für Technologie (KIT)
KIT Scientific Publishing
Straße am Forum 2
D-76131 Karlsruhe

KIT Scientific Publishing is a registered trademark
of Karlsruhe Institute of Technology.
Reprint using the book cover is not allowed.

www.ksp.kit.edu

Print on Demand 2019 – Gedruckt auf FSC-zertifiziertem Papier

ISSN 2191-9925
ISBN 978-3-7315-0967-7
DOI 10.5445/KSP/1000097557

dedicated to
Dilara Alievna Mineeva
*1934, †2002

Contents

1 Introduction

Qubits and Two-Level-Systems (TLS)

Quantum bits (qubits) are elementary blocks of quantum processors that promise a computational power outperforming that of classical computers. Besides trapped ions, and quantum dots formed in semi-conducting heterostructures, superconducting quantum circuits are a prospective realization of qubits convincing by the ease of scalability and a high coherence obeying Moore's law. However, one severe coherence-limiting factor of superconducting qubits are dielectric losses induced at interfaces of the micro-fabricated circuit, that are exposed to oscillating electromagnetic qubit fields. Much effort was spent in the past decades to avoid lossy interfaces [Ste+06; Chu+16] and dilute the qubit fields by clever design and usage of vast qubit footprints [Wan+15; Gam+17]. However the circuit geometry optimization seems to have reached an end, and the necessity grows to step back from quantifying dielectric losses by a uniform loss tangent [OCo+08], and to investigate the microscopic origin of loss: the Two-Level-Systems (TLS).

Generally speaking, TLS are gigahertz resonances known since the 70s, detectable in solids at low temperatures. Their microscopic origin is unclear while the most common and competing models derive the TLS formation from tunneling atoms in amorphous solids [Phi72; AHV72], trapped electrons [FI06; Cho+09; Aga+13], impurities [Mar+05; Sha+10; Gor+14; Hol+13] or adsorbates [Kum+16; Gra+17]. Despite the different physical entities, these models all describe a TLS as a virtual particle trapped in a double-well potential. If a TLS is charged, it possesses an electric dipole moment and may couple to electric fields, and if it has a spin, it correspondingly responds to magnetic fields. An ensemble of TLS with broadly distributed resonance frequencies determines thermal and acoustic properties of the hosting solid at low temperatures and may also dictate its dielectric and magnetic properties. Especially, coupled to dissipative channels such as phonons, electrons or magnons, TLS contribute to corresponding losses in the cold solid.

Since first experimental realizations of qubits in the 2000s [Sim+04; Kim+08], TLS are a hot topic in the quantum computation community. Besides BCS quasiparticles [Cat+11; Pop+14], TLS are known as a main and yet unsolved reason for energy loss and decoherence in qubits but also in superconducting microwave resonators typically used for readout of the qubit quantum state. A common image is that TLS couple via their electric dipole moment to quantum circuits in a dissipative [Mar+05] and dispersive [Cap+18] manner. In the first case, they may resonantly couple to the qubit turning it inoperable in certain spectral regions [Mar+05]. Further, they may absorb energy stored in the circuit, thus limiting the device energy relaxation time [Bar+13] as well as dominate relaxation rate fluctuations in time and frequency [Mül+14; Kli+18]. Second, TLS may undergo spectral diffusion [Mat+16] or telegraphic switching [Mei+18] leading to resonance frequency fluctuations [FI15; Bre+17] of the quantum circuit, thus dephasing it from the measurement apparatus.

In a recent work, phase noise in microwave resonators based on high-kinetic inductance thin wires [Sue+18] was proposed to stem from critical-current fluctuations induced by thermally fluctuating TLS, which could be a new decoherence source in kinetic inductance qubit devices [Gr18]. However, there is no experimental evidence for TLS-induced critical-current fluctuations in Josephson junction-based qubits which are the main qubit-species. Surface spins of adsorbates have been shown to induce charge [Gra+18] and flux noise [Kum+16; Gra+17], while no magnetic TLS resonantly coupled to a qubit have been identified yet.

Independently of the precise coupling mechanism, TLS limit the fidelity of one- and two-qubit gates, as well as the scalability. For this reason, detailed research on the TLS nature is required to deeper understand how TLS occur in the superconducting circuits and how to reduce their number or minimize their impact on the circuit quality. The TLS research at KIT is twofold, so is the work presented here. On the one side, a TLS-sensor architecture has been developed that opens up possibilities for basic research of TLS hosted in thin film and bulk dielectrics, as well as at circuit surfaces. On the other side, methods have been developed to preform TLS experiments with ready-made qubit samples to trace out different TLS locations in the circuit and herewith indicate weak-points of a given device regarding TLS-related losses.

TLS-sensors

We call the TLS-sensor a microwave quantum circuit equipped with a plate or coplanar capacitor containing a dielectric of interest. Hereby, a small distance below 50 nm between the capacitor electrodes is required to reach strong electric fields induced by the circuit, and thus to notably couple to TLS electric dipole moments. A first resonator-based sensor was realized from a lumped element circuit [Sar+16] in the research group of K. D. Osborn, allowing for resonant detection of individual TLS hosted in a thin film dielectric, and their tuning by dc-electric fields induced with on-chip electrodes. At KIT, in the frame of the project led by J. Lisenfeld (DFG Grant LI2446), two straight-forward realizations of TLS-sensors based on resonators were studied. First, the well-elaborated technique to tune TLS by elastic strain transferred from a piezo actuator [Lis+15] was applied to a coplanar wave guide resonator terminated by a small sample capacitor [Bre+17]. Second, a TLS-sensor of equal architecture as in the work [Sar+16] was realized, that allowed TLS tuning by elastic and electric fields [Mat+19].

Realizing TLS-sensors based on tunable resonators would increase the spectral range for TLS sensing while the TLS quantum state could be dispersively detected via the resonator. This would allow one to explore coherence of individual TLS hosted in the sample dielectric, elucidating their interaction with the environment, and giving hints on their microscopic nature. On the other hand, TLS-sensors could be built from qubits that are more sensitive to TLS and offer a wide tunability range. Since at KIT, efficient and fast techniques for detection and coherent readout of TLS employing a superconducting qubit already exist [Coo+04; Lis+10], we decided to build a qubit-based sensor. In this work (DFG Grant LI2446), a TLS-sensor derived from an Xmon [Bar+13] qubit architecture has been developed and fabricated. The detailed working principle is explained in Sec. 2.3.2 while micro-fabrication methods, and associated problems and solutions, are presented in Sec. 3.2 and Chap. 5, respectively. In Sec. 4.1, the experimental proof of principle for a first qubit-based TLS-sensor is reported that opens up the possibility to study TLS in thin and thick dielectric films as well as TLS hosted on film surfaces, i.e., where decoherence generating TLS are suspected.

TLS experiments in ready-made qubits
Investigating TLS via their response to globally applied dc-electric fields, in unchanged superconducting qubits is a non-invasive method to point out regions in the given sample, where TLS contribute most to losses, and to find out which micro-fabrication steps lead to their formation. Hereby, a set of external degrees of freedom is useful to observe TLS behavior in a controlled environment, and draw conclusions about their origin.

Two experiments with Xmon-type transmon qubits [Bar+13] shipped from UCSB, USA were performed in this part of the work. A sample holder was developed that allows one to control elastic strain of the microchip, and the dc-electric field at the sample surface, that is generated by global electrodes surrounding the chip. From tunability characteristics of detected TLS by external fields, conclusions about their possible locations can be drawn. TLS residing in the qubit Josephson junction, e.g., do not couple to the global electrodes while being strain-tunable, as traced out in the first experiment. Or, by comparing the simulated and the measured TLS tunability by each of the global electrodes, one can extract possible TLS locations at qubit film edges. For example, TLS residing at the metal-vacuum interface couple stronger to the global electrode placed above the chip than TLS hosted at the substrate-metal interface, while it is the opposite case regarding the electrode underneath the chip.

After introductory basics of the transmon qubit and TLS in the respective sections 2.1 and 2.2, details are provided on the first and second experiments in Sec. 2.3.5 and 2.3.4. The cryogenic and microwave setups are introduced in Sec. 3.1 while the sample holder is described in Sec. 3.3. The dc-electric fields generated by global electrodes were simulated for interpretation of the experiments. The underlying method and results are reported in Sec. 4.2.1 and 4.2.2, respectively. The method to identify TLS residing inside the Josephson junction are presented in Sec. 4.2.4, verifying that large-area stray Josephson junctions which are a side-effect of the shadow evaporation fabrication technique essentially contribute to dielectric losses. Finally, the TLS location distribution along the qubit film edge profile, and the underlying method that exploits the interplay of both global electrodes are presented in Sec. 4.2.5. The results suggest that about half of all detectable TLS reside at very surfaces of the circuit.

2 Theory and motivations

2.1 Transmon/Xmon basics

2.1.1 Superconductivity and Josephson relations

In 1911, H. Kammerlingh Onnes reported an abrupt resistance drop of mercury down to a non-detectable level, when it was cooled below 4.2 K. He called this material state superconducting. A number of materials undergo such second order phase transition at a material-dependent critical temperature T_c. In contrast to a hypothetical ideal conductor that screens varying magnetic fields, a superconductor additionally expels constant magnetic field once cooled below T_c. This ideal diamagnetism and zero ohmic resistance have been explained in the microscopic theory of superconductivity by Bardeen, Cooper and Schrieffer (BCS) in 1957. The basic idea is that a weak attractive potential between conduction electrons is mediated by the lattice of the solid. At zero temperature, this attraction may cause the Fermi statistics to break down and a Bose condensate to emerge. This is possible since electrons form Cooper pairs which have zero total spin, and obey bosonic statistics and may condensate into a common ground state. The BCS ground state is protected from thermal excitations by an energy gap $\Delta_s(T = 0\,\mathrm{K}) = 1.764\,k_B T_c$ that vanishes for temperatures above T_c so that the normal state is recreated.

In the theory by V. Ginzburg and L. Landau, which was formulated phenomenologically in 1950 and derived from the BCS theory in 1959, the spatial variation of the superconducting state is taken into account. The superfluid condensate is described with a macroscopic wave function $\psi = \sqrt{n(\boldsymbol{r})}\exp\{i\phi(\boldsymbol{r})\}$ that serves as the complex phase transition order parameter. Hereby, $\phi(\boldsymbol{r})$ is the wave function phase, and $n(\boldsymbol{r})$ the Cooper pair density that is the conjugate degree of freedom to ϕ. An important constant introduced in this theory is the coherence length ξ describing the length scale of spatial fluctuations of ψ. Further, ξ describes how far the superconducting wave function penetrates into an adjacent normal metal (N) or insulator (I). Especially, two superconductors separated by a thin N or I layer, or a constriction, may show interference of their wave functions, as analyzed by B. Josephson. He predicted in 1962 the existence of a supercurrent flowing through such a weak link at zero voltage:

$$I_s = I_c \sin(\varphi), \tag{2.1}$$

which is known as the first Josephson relation. Here, φ is the phase difference between the wave functions of both superconductors, and I_c is the maximum supercurrent sustainable by the weak link that is called Josephson junction (JJ). In this work, as typically, JJs are formed of two micro-fabricated aluminum electrodes separated by a thin layer of insulating AlO_x barrier as described in Sec. 3.2.2.

The second Josephson equation expresses the phase drop dynamics when a dc-voltage V is maintained across a JJ:

$$\frac{\mathrm{d}\varphi}{\mathrm{d}t} = \frac{2e}{\hbar}V. \tag{2.2}$$

When applying a current I, a voltage V is induced in the junction. Identifying such behavior with that of a linear inductor $V = -L \cdot \mathrm{d}I/\mathrm{d}t$, and regarding the Josephson relations, the junction inductance takes the form

$$L_\mathrm{J} = \frac{\Phi_0}{2\pi I_\mathrm{c}\cos(\varphi)}. \tag{2.3}$$

Here, $\Phi_0 = \mathrm{h}/2e$ is the magnetic flux quantum in superconductors, h the Planck constant, and e the elementary charge. The Josephson inductance L_J is non-linear since it is dependent on $\cos(\varphi)$. It is a kinetic inductance that stems from the inertia of tunneling Cooper pairs carrying the supercurrent between the junction electrodes. The Josephson junction non-linearity is the main building block of solid-state quantum bits (qubits) realized by superconducting circuits. A qubit is a system with quantized energy states and anharmonic level spacing, which allows for direct and coherent driving of two selected quantum states that mostly are the ground and the first excited state, and that are interpreted as the logical states $|0\rangle$ and $|1\rangle$.

2.1.2 Transmon Hamiltonian

The transmon qubit has been derived from the single Cooper-pair box (SCB) by Koch et al. in a pioneering work [Koc+07] in 2007. The first experimental realization was published in 2008 by Schreier et al. [Sch+08]. As shown in the circuit diagram of Fig. 2.1 (a), it consists of a Josephson junction (JJ) shunted with a large capacitor C_B. Since the current through the JJ and through the capacitor is conserved, we can find the equation of motion in terms of the phase drop φ across the junction using Eqs. (2.1) and (2.2):

$$I_\mathrm{c}\sin(\varphi) + \frac{\hbar}{2e}C_\Sigma\ddot{\varphi} = 0, \tag{2.4}$$

where I_c is the junction critical current, and C_Σ is the sum of C_B and the JJ intrinsic capacitance C_J. From the Euler-Lagrange equation we then derive the Lagrange function

$$\mathscr{L} = \frac{C_\Sigma}{2}\left(\frac{\hbar}{2e}\dot{\varphi}\right)^2 + E_\mathrm{J}\cos(\varphi), \tag{2.5}$$

where the Josephson energy E_J is defined as

$$E_\mathrm{J} = \frac{I_\mathrm{c}\hbar}{2e}. \tag{2.6}$$

The charge stored in the capacitor is the conjugate degree of freedom to the phase φ, and is defined as

$$q \equiv \frac{2e}{\hbar}\frac{\partial \mathscr{L}}{\partial \dot{\varphi}}, \tag{2.7}$$

that fulfills the commutation relation $[q, \varphi] = \mathrm{i}2e$. The Hamiltonian is derived via the Legendre transformation:

$$H = \dot{\varphi}\frac{\partial}{\partial \dot{\varphi}}\mathscr{L} - \mathscr{L} \tag{2.8}$$

$$= \frac{1}{2C_\Sigma}q^2 - E_\mathrm{J}\cos(\varphi) \tag{2.9}$$

$$\equiv 4E_\mathrm{C}n^2 - E_\mathrm{J}\cos(\varphi), \tag{2.10}$$

where $n = q/2e$ is defined as the number of Cooper-pairs and E_C is the so-called charging energy:

$$E_\mathrm{C} = \frac{e^2}{2C_\Sigma}. \tag{2.11}$$

In contrast to the SCB, the transmon is operated in the regime of a high $E_\mathrm{J}/E_\mathrm{C}$ ratio of about 100. The transmon quantized energy levels can be understood as those of a virtual phase particle trapped in the sinusoidal Josephson potential $-E_\mathrm{J}\cos(\varphi)$. As derived in the work [Koc+07], the transition energy from the ground to first excited state (logical states $|0\rangle$ and $|1\rangle$) is $E_{01} = \sqrt{8E_\mathrm{C}E_\mathrm{J}} - E_\mathrm{C}$, where the energy correction $-E_\mathrm{C}$ is due to the potential anharmonicity.

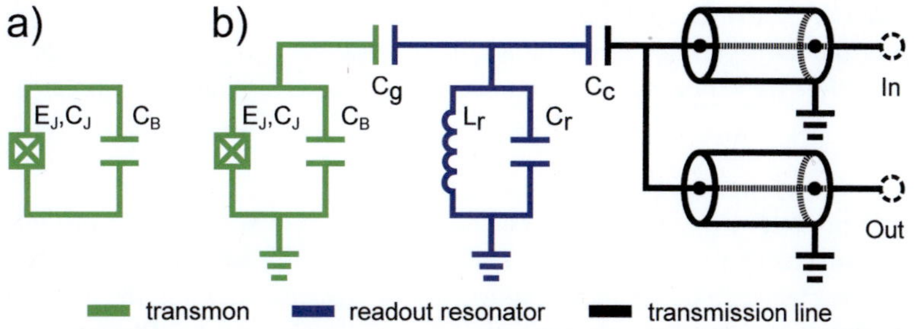

Figure 2.1: a A qubit defined by a Josephson junction (JJ) shunted with a capacitor C_B. It is called a transmon if its designed parameters allow the regime $E_\mathrm{J}/E_\mathrm{C} \sim 100$, where the charging energy is defined as $E_\mathrm{C} \equiv e^2/2C_\Sigma$, and C_Σ is the sum of C_B and the intrinsic junction capacitance $C_\mathbf{J}$. Further, $E_\mathrm{J} = I_\mathrm{c}\hbar/2e$ is the Josephson energy, and derived from the JJ critical current $I_\mathbf{c}$. In this regime, typically C_B is $\sim 100\,\mathrm{fF}$ and about 30 times larger than $C_\mathbf{J}$. The underlying Hamiltonian is shown in Eq. (2.10) [Koc+07]. **b** Xmon qubit coupled to its readout circuitry consisting of a harmonic oscillator (blue) and a transmission line (black colored circuit). In this constellation, the charging energy slightly changes due to dressed $C_\Sigma = C_\mathrm{B} + C_\mathrm{J} + C_\mathrm{g}$. The cavity resonance $\omega_\mathrm{r} = 1/\sqrt{L_\mathrm{r}C_\mathrm{r}}$ is given by its inductance L_r and capacitance C_r while C_C determines the resonator loaded quality factor. The transmon is not gate-tunable, which is an important property for the experiments in this work (see Sec. 2.3).

In this work, it is necessary to tune the qubit in a wide frequency range, in order to resonantly detect material defects introduced in Sec. 2.2. Therefor, the single Josephson junction is replaced by a split-junction that consists of two parallel JJs with critical currents I_{c1} and I_{c2}. An applied magnetic flux Φ to such a dc-SQUID modulates the maximum critical current and herewith the transition energy

$$E_{01}(\Phi) = \sqrt{8E_C E_{J\Sigma} \left|\cos\left(\frac{\pi\Phi}{\Phi_0}\right)\right| \sqrt{1 + d^2 \tan^2\left(\frac{\pi\Phi}{\Phi_0}\right)}} - E_C, \tag{2.12}$$

where $E_{J\Sigma} = (I_{c1} + I_{c2})\hbar/2e \equiv I_c\hbar/2e$ is the maximum SQUID Josephson energy, and d is the junction asymmetry defined as $d = (I_{c1} - I_{c2})/(I_{c1} + I_{c2})$ which is designed as zero for qubits fabricated in this work. It is practical to express the transmon resonance frequency $f_{01} = E_{01}/\mathrm{h}$ at the qubit sweet spot ($\Phi = 0$) in terms of R_n that is the dc-SQUID ohmic resistance measurable at room temperature:

$$f_{01} = \frac{1}{\mathrm{h}}\sqrt{\frac{E_C \mathrm{h}\Delta_s(T)}{e^2 R_n}\tanh\left(\frac{\Delta_s(T)}{2\mathrm{k_B}T}\right)} - \frac{E_C}{\mathrm{h}} \tag{2.13}$$

$$\approx \frac{1}{\mathrm{h}}\sqrt{\frac{E_C \mathrm{h}\Delta_s(T)}{e^2 R_n}} - \frac{E_C}{\mathrm{h}}. \tag{2.14}$$

Here, the Ambegaokar-Baratoff relation

$$I_c = \frac{\pi\Delta_s(T)}{2eR_n}\tanh\left(\frac{\Delta_s(T)}{2\mathrm{k_B}T}\right) \tag{2.15}$$

$$\tag{2.16}$$

was used, where I_c is the dc-SQUID critical current. The expression (2.13) simplifies when accounting for the low operation temperature of the qubit $T \approx 30\,\mathrm{mK}$ and the superconducting gap $\Delta_s(T) \approx 2.78 \times 10^{-23}\,\mathrm{J} \approx 2\,\mathrm{K}$ of Aluminium which is used as material of junction electrodes in this work.

The qubit type investigated in this work is called Xmon [Bar+13]. While the original transmon architecture consists of two floating electrodes separated by a JJ, Xmons are galvanically connected at one electrode to the surrounding ground plane as shown in the circuit diagram in Fig. 2.1 (b). This does not affect the transmon physics, however it facilitates the experiments on material defects in the qubit circuit, as it will be explained in section 2.3.

2.1.3 Dispersive readout of transmon qubits

Since the 1960's, cavity quantum electrodynamics (CQED) studies the interaction between atoms and light in a cavity. The atom is understood as a quantum system with quantized energy levels that couple to discrete resonant modes of the cavity as described by the

Jaynes-Cummings Hamiltonian [JC63]. Under certain conditions, the cavity modes experience a dispersive or AC Stark shift dependent on the atom excitation that can be employed for detection of the atom quantum state. In 2004, Blais et al. [Bla+04] proposed to employ this readout scheme to an artificial atom realized by a superconducting qubit coupled to a readout microwave resonator that replaces the optical cavity. In the following, basics of the dispersive readout of a transmon qubit [Koc+07; Ree+10] are presented.

The dispersive readout first requires the strong coupling regime defined by $g_\mathrm{q} \gg \kappa, \Gamma_{1,\mathrm{Q}}$, where g_q is the coupling strength of qubit and resonator while κ and $\Gamma_{1,\mathrm{Q}}$ are the cavity and qubit energy relaxation rates, respectively. Second, the frequency detuning between the qubit and resonator has to exceed g_q. Fig. 2.1 (b) contains a diagram of an Xmon-type transmon (green) coupled to its readout circuitry that consists of a readout resonator (blue) and a transmission line (black). The unperturbed transmon Hamiltonian in Eq. (2.10) does not change in this circuit except for the definition the qubit total capacitance

$$C_\Sigma = C_\mathrm{B} + C_\mathrm{J} + C_\mathrm{g}. \tag{2.17}$$

In contrast to the model presented in [Bla+04], the transmon is not a pure qubit but rather an anharmonic oscillator that obeys the generalized Jaynes-Cummings Hamiltonian [Koc+07]:

$$H_\mathrm{JC} = \hbar \sum_i \omega_i |i\rangle\langle i| + \hbar\omega_\mathrm{r} \hat{a}^\dagger \hat{a} + \hbar \sum_{i,j} g_{i,j} |i\rangle\langle j| (\hat{a}^\dagger + \hat{a}), \tag{2.18}$$

where $\hbar\omega_i$ and $\hbar\omega_\mathrm{r}$ denote the transmon and resonator energy levels, respectively. $\hat{a}$ is the resonator annihilation operator and $g_{i,j}$ is the coupling energy defined by the coupling capacitance C_g. The equation (2.18) expresses that virtual transitions through higher transmon levels have to be taken into account in the calculations of the dispersive shift. After applying a rotating wave approximation and reducing the Hilbert space of the transmon to its lowest two states [Koc+07], two distinct dispersive shifts turn out to be of importance. The first one is detectable while the qubit remains in its ground state. It is the frequency shift $\nu^{(0)}$ from the resonator few-photon state to its many-photon bright state, in which it is decoupled from the transmon [Ree+10]:

$$\nu^{(0)} = \frac{g_\mathrm{q}^2}{\Delta}. \tag{2.19}$$

Here, the detuning is defined as $\Delta = \omega_{01} - \omega_r$ ($\omega_{01} = 2\pi f_{01} = \omega_1 - \omega_0$), and the coupling energy as $\hbar g_\mathrm{q} \equiv \hbar g_{01}$. The shift $\nu^{(0)}$ is useful to measure g_q via continuous spectroscopy. The second dispersive shift ν is used for qubit readout in frequency-domain and time-domain setups. It is the difference between the resonator few-photon states when the qubit is in the ground or first excited state, respectively:

$$\nu = \frac{2 g_\mathrm{q}^2 \alpha}{\Delta(\Delta - \alpha)}. \tag{2.20}$$

Here, α denotes the transmon anharmonicity defined as

$$\alpha \equiv \omega_{12} - \omega_{01} \approx -E_\mathrm{C}/\hbar, \tag{2.21}$$

which is typically $-2\pi \cdot 200\,\mathrm{MHz}$.

2.2 Basics of Two-Level-Systems (TLS)

Until the 1970s, it has been argued that thermal and acoustic properties of solids at low temperatures are determined solely by quantized vibrational modes of the lattice (phonons). In this case, no difference between amorphous and crystalline solids was expected due to the long wave length of phonons at low temperatures, which exceeds by orders of magnitude the inter-atomic distance. This picture however did not explain the contemporary observations of thermal and acoustic properties of cold glasses, which differed substantially from the Debye model predictions. The unexpected high heat capacity and low thermal conductance, e.g., suggested additional degrees of freedom present in disordered solids, which at low temperatures may absorb (or store) energy and scatter phonons.

The concept of tunneling atoms is known since 1930s while adopted to amorphous solids by Phillips and Anderson in 1972 [Phi72; AHV72] to derive thermal and acoustic properties of cold glasses from ensembles of "tunneling states" or "Two-Level-Systems" (TLS). The basic idea of the so-called standard tunnel model is that single or few atoms may find two energetically almost equivalent metastable states in the glass lattice, as illustrated in Fig. 2.2 (a). At low temperatures, these states may be accessible through coherent tunneling, which forms a single TLS associated with a certain transition frequency. Regarding the fact, that TLS may also occur in crystals from defects or from impurities, later on, Black [Bla81] summarized that the "glassiness [..] is neither totally sufficient nor absolutely necessary for the occurrence of TLS". However, it is clear that the local disorder inherent to glasses is necessary for a broad distribution of TLS transition frequencies observable in superconducting quantum circuits [Mar+05].

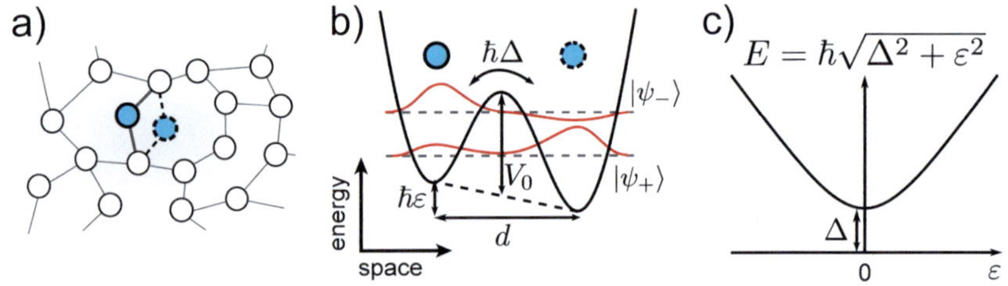

Figure 2.2: **a** Illustration of a Two-Level-System (TLS) formed by one atom tunneling between two metastable states in an amorphous solid, according to the model that was developed by Phillips [Phi72] and Anderson et al. [AHV72] to explain acoustic and thermal properties of cold glasses. **b** Besides structural defects in solids, TLS can possibly be formed by a variety of microscopic systems like trapped electrons [FI06; Cho+09; Aga+13], impurities [Mar+05; Sha+10; Gor+14; Hol+13] or adsorbates [Kum+16; Gra+17]. However, most common mechanisms can be modeled as a virtual particle trapped in a one-dimensional double-well potential. In the special example shown in (a) and (b), the potential minima represent the two metastable locations of an atom in an amorphous matrix. **c** The TLS transition frequency derived from the double-well problem evolves hyperbolically versus varying potential asymmetry ε. This well-known behavior is routinely observed in the experiments of this work, where static electric fields and elastic strain are applied to the sample to tune individual TLS resonance frequencies.

Along with the standard tunneling model (STM) by Phillips and Anderson, a variety of other models exist that derive TLS from impurities [Mar+05; Sha+10; Gor+14; Hol+13], trapped electrons [FI06; Aga+13] or adsorbates [Kum+16; Gra+17], as summarized in the review by Müller et al. [MCJ17] on TLS in quantum devices. However, disregarding the physical entity all these models boil down to the STM Hamiltonian

$$\widetilde{H}_{\mathrm{TLS}} = \frac{\hbar}{2}\begin{pmatrix} \varepsilon & \Delta \\ \Delta & -\varepsilon \end{pmatrix}, \tag{2.22}$$

that describes a TLS as a virtual particle trapped in a one-dimensional double-well potential illustrated in Fig. 2.2 (b). Here, $\hbar\varepsilon$ is the asymmetry energy and $\hbar\Delta$ the coupling energy between the lowest bound states $|\mathrm{L}\rangle$ and $|\mathrm{R}\rangle$ in each potential minimum that depends on the barrier height V_0 and the spacing d. While the potential shape depends on the underlying model and is uncertain in amorphous systems, in all common models Δ depends exponentially on V_0 and d. A suitable approach is to describe the double-well potential with two overlapping harmonic potentials, so that

$$\Delta = \hbar\omega_0 \exp\left\{-\frac{d}{\hbar}\sqrt{2mV_0}\right\}, \tag{2.23}$$

where ω_0 is approximately the average frequency of the localized states $|\mathrm{L}\rangle$ and $|\mathrm{R}\rangle$ in each potential minimum [Phi87]. In the diagonalized basis, Eq. (2.22) reads as

$$H_{\mathrm{TLS}} = \frac{E}{2}\sigma_z, \tag{2.24}$$

and its eigenstates are

$$|\psi_-\rangle = \cos\left(\frac{\theta}{2}\right)|\mathrm{L}\rangle - \sin\left(\frac{\theta}{2}\right)|\mathrm{R}\rangle, \tag{2.25}$$

$$|\psi_+\rangle = \sin\left(\frac{\theta}{2}\right)|\mathrm{L}\rangle + \cos\left(\frac{\theta}{2}\right)|\mathrm{R}\rangle, \tag{2.26}$$

where the mixing angle is defined as $\theta = \mathrm{atan}(\Delta/\varepsilon)$ and σ_z is the diagonal Pauli matrix. The TLS transition energy E reads as

$$E = E_- - E_+ = \hbar\sqrt{\Delta^2 + \varepsilon^2}, \tag{2.27}$$

and depends hyperbolically on ε, as shown in Fig. 2.2 (c). For a symmetric TLS ($\varepsilon = 0$), the virtual particle is maximally delocalized between the potential minima, while for $\varepsilon \gg \Delta$ it is trapped in one minimum.

The TLS Hamiltonian in Eq. (2.22) reminds of a spin $1/2$ exposed to a magnetic field. In contrast to this, the TLS tunnel energy is an intrinsic parameter and does not change, while the asymmetry energy linearly scales when applying strain or electric fields to the hosting solid [AHV72; Phi72]. The part of the Hamiltonian describing the TLS coupling to external fields is thus diagonal in the localized basis $\{|\mathrm{L}\rangle, |\mathrm{R}\rangle\}$:

$$\widetilde{H}_{\mathrm{int}} = \begin{pmatrix} 1 & 0 \\ 0 & -1 \end{pmatrix} \boldsymbol{dE} + \begin{pmatrix} 1 & 0 \\ 0 & -1 \end{pmatrix} \gamma \xi, \tag{2.28}$$

where $\boldsymbol{E}$ is the electric and ξ the strain/elastic field. The vectorial parameter $\boldsymbol{d}$ is the TLS electric dipole moment defined as $\partial \varepsilon / 2 \partial \boldsymbol{E}$. The analog to the TLS elastic dipole moment is $\gamma \equiv \partial \varepsilon / 2 \partial \xi$ whose tensorial character is averaged over all directions. When applying adiabatically varying fields, the TLS eigenstates will instantaneously diagonalize, with the transition Energy

$$E(\boldsymbol{E}, \xi) = \hbar \sqrt{\Delta^2 + \left(\varepsilon + \frac{2\boldsymbol{dE}}{\hbar} + \frac{2\gamma\xi}{\hbar} \right)^2}, \tag{2.29}$$

where ε is understood as the intrinsic TLS asymmetry. In the experiments of this work, constant electric fields were generated either by on-chip or by global gates, and static strain was transferred to the sample chip using a piezo actuator to tune individual TLS via their electric and elastic dipole moments, respectively.

When considering periodic weak fields, it is helpful to rewrite $\widetilde{H}_{\mathrm{int}}$ in the diagonalized basis $\{|\psi_+\rangle, |\psi_-\rangle\}$:

$$H_{\mathrm{int}} = \left(\frac{\varepsilon}{E} \sigma_z + \frac{\Delta}{E} \sigma_x \right) \boldsymbol{dE} + \left(\frac{\varepsilon}{E} \sigma_z + \frac{\Delta}{E} \sigma_x \right) \gamma \xi, \tag{2.30}$$

where the off-diagonal terms allow for resonant excitation of TLS. Via their deformation potential and the electric dipole moment, TLS couple respectively to local strain generated by phonons and to local electric fields induced by charge carriers like conducting electrons present in metallic glasses, or BCS quasiparticles in superconducting glasses [Bla81] and in qubits [Bil+17]. This constitutes a main energy loss mechanism of TLS. On the other side, TLS are a main decoherence source of superconducting qubits since they may couple to the qubit oscillating fields by their electric dipole moments. As a result, they can resonantly absorb the qubit excitation [Mar+05] or dispersively shift its resonance frequency, as demonstrated for microwave resonators [Cap+18], which are a similar system. The qubit sensitivity to TLS is exploited in this work to resonantly detect individual TLS, as further explained in the following section.

2.3 TLS in this work

This work comprises two types of experiments: exploring TLS in sample dielectrics using TLS-sensors derived from a tailored qubit circuit, and detection of TLS in ready-made qubit samples. Custom-made TLS-sensors offer insights to the TLS nature that are of great interest for the optimization of quantum circuits regarding the TLS-related energy loss and decoherence. Investigation of TLS in state-of-the-art qubit samples enables one to point out the circuit-specific weak-points that contribute the most to TLS-related losses. In Sec. 2.3.1, a method to resonantly detect individual TLS, and an estimation of the TLS detection sensitivity of the qubit are presented. In Sec. 2.3.2, the working principle of a transmon-based TLS-sensor is explained. In Sec. 2.3.4, an experimental method to distinguish TLS hosted in the qubit Josephson junction is presented. In the last section 2.3.5, a technique to determine positions of surface-TLS at the qubit film edges using multiple global dc-electrodes is shown in details.

2.3.1 Swap-spectroscopy and TLS detection sensitivity

All experiments in this work require detection of individual TLS exploiting the qubit tunability. The underlying so-called swap-spectroscopy protocol first demonstrated by Cooper et al. [Coo+04] and well established at KIT [Lis+10] is shown in Fig. 2.3 (a). The qubit is excited by a π-pulse, tuned to a variable probe frequency via a flux pulse for a constant time τ that is comparable to the qubit energy relaxation time $T_{1,\mathrm{Q}}$, and then read out. When the qubit is near resonance with a TLS at a given frequency, its excitation decays faster and a drop of its excited state population is detected. This protocol is repeated for various flux pulse amplitudes to record the TLS resonance frequency distribution in the qubit tunability range. The time-dependent distortions of the rectangular flux pulse due to impedance mismatch-related reflections are corrected as described in the supplementary of the work [Bar+13]. From the qubit coherence time and τ, one can deduce the minimum coupling strength g, for which a TLS is still detectable. In Fig. 2.3 (b) the measurement protocol to deduce the qubit $T_{1,\mathrm{Q}}$ is shown. In Fig. 2.3 (c) the corresponding data is shown that was obtained with Xmon No. 3 from the UCSB sample [Bar+13] without any TLS in resonance. The fit $P_0 \exp\{-t\Gamma_{1,\mathrm{Q}}\}$ is indicated by the continuous black line with $\Gamma_{1,\mathrm{Q}}$ being the inverse $T_{1,\mathrm{Q}}$. When the qubit is near resonance with a TLS, its energy relaxation rate results from the Master equation solution (see Supp. of [Bar+13]):

$$\Gamma_1 = \frac{2g^2\Gamma}{\Gamma^2+\Delta^2} + \Gamma_{1,\mathrm{Q}}, \tag{2.31}$$

where Δ is the detuning between qubit and TLS while $\Gamma = \Gamma_{1,\mathrm{TLS}}/2 + \Gamma_{\phi,\mathrm{TLS}} + \Gamma_{1,\mathrm{Q}}/2 + \Gamma_{\phi,\mathrm{Q}}$. Here, $\Gamma_{1,\mathrm{TLS}}$ is the energy relaxation rate of a TLS and $\Gamma_{\phi,\mathrm{TLS}}$ its pure dephasing rate. The expression (2.31) is valid when $\Gamma_{1,\mathrm{TLS}} > g > \Gamma_{1,\mathrm{Q}}$, which is the case assuming a typical $\Gamma_{1,\mathrm{TLS}}/2 + \Gamma_{\phi,\mathrm{TLS}} \approx 10/\mu\mathrm{s}$ [Bar+13], a typical g of about $1.3/\mu\mathrm{s}$ [Bar+13] and $\Gamma_{1,\mathrm{Q}} < 0.5/\mu\mathrm{s}$ in this work. The minimum detectable g is deduced from a condition that is visualized in Fig. 2.3 (c). Namely, the difference between the free decay $P_0 \exp\{-\tau\Gamma_{1,\mathrm{Q}}\}$ (black line) and the TLS-enhanced decay $P_0 \exp\{-\tau\Gamma_1\}$ (red dashed line) has to exceed the noise level 2σ (red error bar):

$$g_{\min} = \sqrt{-\frac{\Gamma}{2}\left(\frac{1}{\tau}\ln\left(\exp(-\tau\Gamma_{1,\mathrm{Q}}) - \frac{2\sigma}{P_0}\right) + \Gamma_{1,\mathrm{Q}}\right)}. \tag{2.32}$$

Here, σ is the standard deviation of the data points when no π-pulse is applied, which has been observed to be about 0.04 in the experiments performed in this work with UCSB samples. The sample Xmon No. 3 that has $T_{1,\mathrm{Q}} = 7.2\,\mu\mathrm{s}$ can thus resolve TLS at a minimum coupling strength $g/2\pi > 0.07\,\text{MHz}$. Minimizing g_{min} leads to the optimum τ, which gives a rule of thumb in accordance to our experience:

$$\tau \approx 0.8 T_{1,\text{Q}}. \tag{2.33}$$

The here-presented estimation of g_{min} is simplified since it ignores the readout duration of minimum 500 ns. If $T_{1,\text{Q}}$ is comparable with that time scale, which is the case for the first qubit-based TLS-sensor version (see Sec. (4.2)), this method breaks down. Further, when speaking of g_{min} we mean the value deduced from the typical unperturbed qubit $T_{1,\text{Q}}$ disregarding fluctuations in time, typically observed in transmon qubits [Mül+14; Kli+18; Sch+19]. In the experiments of this work, the swap-spectroscopy was repeated for varying external dc-electric and elastic fields which tune individual TLS but do not disturb the qubit. The measurements lasted for days while the qubit frequency was recalibrated each few hours.

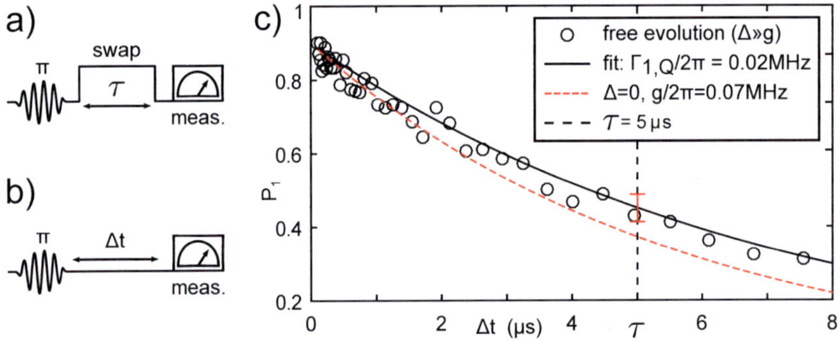

Figure 2.3: **a** The swap-spectroscopy protocol used in all experiments of this work to resonantly detect individual TLS within the frequency tunability range of the qubit. **b** Measurement scheme to deduce the qubit energy relaxation rate $\Gamma_{1,\text{Q}}$. **c** Exemplary data obtained with the UCSB sample Xmon 3 [Bar+13] using the protocol in (b). The black line is the fit to the exponential decay of the qubit excited state population. When the qubit is tuned into resonance with a TLS using the protocol in (a), the theoretical qubit decay constant is given by Eq. (2.31), and the corresponding qubit relaxation is represented by the red dashed line. The difference between both lines is the information indicating the existence of a TLS at a given frequency. Entering the swap duration τ and the standard deviation (red error bar) into Eq. (2.32), the minimum resolvable difference is deducible and herewith the minimum coupling strength g_{min} (see red dashed line in the legend), for which a TLS is still detectable.

2.3.2 An Xmon-based TLS-sensor

The principle of a qubit-derived TLS-sensor has been developed by J. Lisenfeld in the frame of the Deutsche Forschungsgemeinschaft (DFG) Grant LI2446/1. The idea is to shunt the Josephson junction (JJ) of a qubit with a small capacitor containing a sample dielectric. The circuit schematic in Fig. 2.4 (a) shows an Xmon-type qubit equipped with such a sample capacitor. The sample dielectric can be embedded in a trilayer geometry that we call the overlap contact (OV), or deposited on top of a coplanar capacitor called the nano-gap (NG).

We call the TLS hosted in such a capacitor sample-TLS. In the OV, the coupling strength g between the qubit and the sample-TLS has the form

$$\hbar g = \frac{\Delta}{E} d_{\parallel} E_{\text{qub}}^{\text{rms}} = \frac{\Delta}{E} d_{\parallel} \frac{1}{t_{\text{OV}}} \sqrt{\frac{\text{h} f_{01}}{2 C_{\text{qub}}}}, \tag{2.34}$$

where, $d_{\parallel}$ is the TLS dipole moment component that is parallel to the qubit field, and Δ/E is the TLS matrix element of the transversal term in Eq. (2.30). The TLS is maximally coupled to the qubit when its electric dipole moment is parallel to the qubit field, and when it is symmetric ($\varepsilon = 0$, see Eq. (2.27)). The sample dielectric thickness is given by t_{OV} while

$$C_{\text{qub}} = C_{\Sigma} + n C_{\text{OV}} = C_{\Sigma} + n \varepsilon_0 \varepsilon_{\text{r}} \frac{A}{t_{\text{OV}}} \tag{2.35}$$

is the qubit total capacitance comprising the qubit intrinsic capacitance $C_{\Sigma} = C_{\text{B}} + C_{\text{J}} + C_{\text{g}}$ and the n sample capacitors. C_{B} is the capacitance of the qubit island to ground, C_{J} is the JJ capacitance, and C_{g} is the qubit coupling capacitance to the readout resonator. A is the area of the overlap contact. The root mean square of the qubit electric field is calculated regarding the qubit as a harmonic oscillator in the ground state:

$$E_{\text{qub}}^{\text{rms}} = \frac{U_{\text{qub}}^{\text{rms}}}{t_{\text{OV}}} = \frac{1}{t_{\text{OV}}} \sqrt{\frac{1}{T} \int_0^T U_{\text{qub}}(t)^2 dt} \tag{2.36}$$

$$= \frac{1}{t_{\text{OV}}} \sqrt{\frac{\text{h} f_{01}}{2 C_{\text{qub}}}}, \tag{2.37}$$

where T is the oscillation time period and $U_{\text{qub}}(t) = \sqrt{\text{h} f_{01}/C_{\text{qub}}} \sin(2\pi f_{01} t)$. For the nano-gap with a gap width of t_{OV}, Eq. (2.34) gives only the approximate maximum g for a given $d_{\parallel}$.

In the detector architecture shown in Fig. 2.4 (a), one could try to distinguish sample-TLS, and the intrinsic TLS hosted at the qubit surface (surface-TLS) and in the qubit JJ (junction-TLS) by comparing the measured g. This is however unfeasible without statistical uncertainties due to random orientation of the electric TLS dipole moment. The modified architecture presented in Fig. 2.4 (b) allows one to tune the sample-TLS with an electric field by applying a dc Voltage V_{DC} to the sample capacitor (blue circuit part). Due to the transmon regime, Cooper pairs can tunnel through the JJ and equilibrate the dc potential drop between the qubit island (green) and the ground plane (black). Thus, the junction-TLS and the surface-TLS do not experience any electric field change since no voltage drop is induced within the JJ or at the surface of the qubit island. This setup thus makes the sample-TLS distinguishable from the intrinsic ones with a negligibly small error: In a small region of $\sim 100\,\text{nm}$ outside of the sample capacitors, surface-TLS may experience the stray fields from the biased OV electrode. Their average density is however assumed about $0.1\,/(\mu\text{m})^2$.

The large filter capacitor ($C_{\text{f}} \sim 250\,\text{fF}$) indicated in Fig. 2.4 (b) serves in this architecture as a dc-break between the qubit island and the ground plane while acting as a low pass for the dc voltage source, protecting the qubit from noise. Since the filter is placed in series with the sample capacitor (~ 0.2 to $1.1\,\mathrm{fF}$), more than 99.6% of the qubit-induced voltage drops across the sample capacitor so that we can be sure the detected sample-TLS not to stem from the filter. For the same reason, the filter practically does not participate in dielectric losses and is disregarded in the following participation ratio discussion. In the experiment, when a TLS-sensor is equipped with two sample capacitors in parallel, each of them has a filter and can be individually biased with a dc voltage.

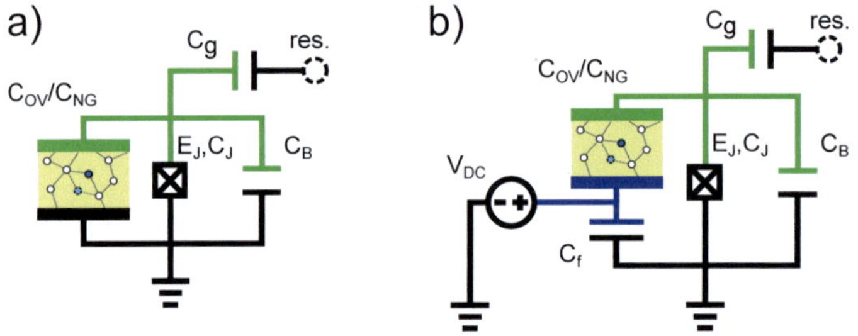

Figure 2.4: a Simple architecture of a TLS-sensor based on an Xmon qubit introduced in Sec. 2.1.2. The readout circuitry shown in Fig. 2.1 (b) is omitted here. The small capacitor containing the sample dielectric (yellow) can be realized as a plate- (C_{OV}) or a coplanar (C_{NG}) capacitor, both with a typical electrode distance of $\sim 50\,\mathrm{nm}$. The simple detector architecture however does not allow an errorless distinction between TLS hosted in the sample dielectric (sample-TLS) and the intrinsic ones. **b** The TLS-sensor architecture used in this work enables one to tune the sample-TLS by applying a dc voltage V_{DC} to the sample capacitor (blue). Hereby, the qubit island (green) is at the same dc potential as the ground plane due to the transmon regime so that intrinsic TLS hosted in the qubit Josephson junction (JJ) or at qubit interfaces do not respond to V_{DC}. The large filter capacitor C_{f} serves as a dc-break in the qubit circuit and as a low pass filter protecting the qubit from the voltage source noise. In our experiments, we used detectors with one and two sample capacitors placed in parallel to the qubit JJ.

In an overlap contact, t_{OV} determines the typical coupling strength of the sample-TLS, and its capacitance C_{OV} determines the dielectric losses. The energy relaxation rate of a qubit equipped with n OVs can be decomposed into distinct contributors [Wan+15]:

$$\Gamma_1 = 2\pi f_{01} \sum_i \frac{p_i}{Q_i} + \widetilde{\Gamma}_0 \tag{2.38}$$

$$\equiv 2\pi f_{01} \frac{p_{\mathrm{OV}}(n)}{Q_{\mathrm{OV}}} + \Gamma_0, \tag{2.39}$$

where the sum goes over all capacitive elements of the qubit circuit like the interfaces, JJs and the sample capacitors. The individual participation p_i is defined as the ratio of the charging energy stored in the element i divided by the total charging energy. Q_i is the quality factor of the ith element. $\widetilde{\Gamma}_0$ denotes all other losses in the circuit like BCS quasiparticle [Cat+11] or vortex-related ones [Son+09]. In Eq. (2.39), the contribution by the OVs is emphasized while Γ_0 comprises the residual losses. Here, Q_{OV} is the inverse loss tangent of the sample dielectric while

$$p_{\mathrm{OV}}(n) = \frac{nC_{\mathrm{OV}}}{C_{\mathrm{qub}}} \tag{2.40}$$

is the participation ratio of n OVs. Thus, knowing or making reasonable assumptions about the unperturbed losses Γ_0 and having in mind the target Γ_1, one can determine the needed overlap contact's area A. When dealing with nano-gaps, p_{NG} results from p_{OV} by replacing C_{OV} by C_{NG}. Since in this work only experiments with OV-equipped sensors are reported, the NG is not handled in further details.

Using the TLS-sensor equipped with OVs, one can precisely determine $d_{\parallel}$ from the tunability characteristics of sample-TLS measured in V_{DC} sweeps. The TLS transition frequency introduced in Sec. 2.2 has the following dependency:

$$\omega_{\mathrm{TLS}}(V_{\mathrm{p}}, V_{\mathrm{DC}}) = \sqrt{\Delta^2 + (\varepsilon + \gamma_{\mathrm{p}} V_{\mathrm{p}} + \gamma_{\mathrm{DC}} V_{\mathrm{DC}})^2}. \tag{2.41}$$

Since we exploit the long-established technique to transfer elastic strain onto the sample (see Sec. 3.3.1), TLS may be tuned via strain. Thus, $\hbar\varepsilon$ is the sum of the intrinsic TLS asymmetry energy and that added by pre-strain from to the piezo actuator which has a different thermal contraction coefficient compared to the sample housing. γ_{p} is the individual TLS sensitivity to the elastic field comprising the TLS position in the circuit and its tensorial deformation potential. Accordingly, γ_{p} is a fit factor to $\omega_{\mathrm{TLS}}(V_{\mathrm{p}}, const)$. Further, γ_{DC} is the individual TLS sensitivity to the applied dc voltage V_{DC} that depends on the orientation of the individual TLS electric dipole moment relative to the electric field E_{OV}. Using the identity

$$\hbar\gamma_{\mathrm{DC}} V_{\mathrm{DC}} = 2d_{\parallel} E_{\mathrm{OV}} = 2d_{\parallel} \frac{V_{\mathrm{DC}}}{t_{\mathrm{OV}}}, \tag{2.42}$$

we can deduce $d_{\parallel}$ from the fitted γ_{DC} to $\omega_{\mathrm{TLS}}(const, V_{\mathrm{DC}})$:

$$d_{\parallel} = \frac{\hbar\gamma_{\mathrm{DC}} t_{\mathrm{OV}}}{2}. \tag{2.43}$$

Since the qubit field direction within the sample dielectric coincides with that induced by V_{DC}, we can deduce the TLS-qubit coupling strength g from $d_{\parallel}$ using Eq. (2.34). The experimental results on TLS detection in amorphous and insulating AlO_x are presented in Sec. 4.1. Finally, it is important to emphasize that the Xmon architecture fits best for a TLS-sensor that allows for exact identification of sample-TLS. In the next section we show that TLS-sensors based on a floating transmon do not allow this functionality.

2.3.3 TLS-sensor based on a floating transmon qubit

There is a bunch of possibilities to build a TLS-sensor based on a floating transmon qubit that allows one to tune the sample-TLS by dc-electric fields. However, none of them seems to fully circumvent unwanted tuning of TLS hosted outside of the sample dielectric due to the floating qubit topology, which makes sample-TLS and intrinsic TLS not fully distinguishable. Fig. 2.5 contains a circuit diagram and a design sketch of an exemplary TLS-sensor that is similarly built as the Xmon-based detector presented in the previous section. Some stray capacitances are disregarded here since they are not necessary to point out the basic problem of the sensor type derived from a floating transmon. When applying a dc voltage V_{DC} to the first sample capacitor, the dc potential drop between the qubit islands is zero due to the transmon regime, while each sample capacitor experiences a voltage V_i ($i \in \{1,2\}$), were $V_1 + V_2 = V_{DC}$. Both islands will have a non-zero potential V_2 relative to the ground plane, that is proportional to V_{DC}. Thus, intrinsic TLS which reside at qubit film edges may be tunable by the electric field induced between qubit islands and the ground plane. It is possible to apply voltages to both sample capacitors and find a regime where the qubit island has zero potential, this however is a way more complicated experiment than that executed in this work on Xmon-based detectors. Another transmon-based detector architecture can be derived from the circuit design used in the work of Sarabi et al. [Sar+16] by replacing the resonator geometric inductance by a Josephson junction shunted with a large capacitor. In this design, exactly the same problem of non-zero field between qubit islands and the ground plane would arise.

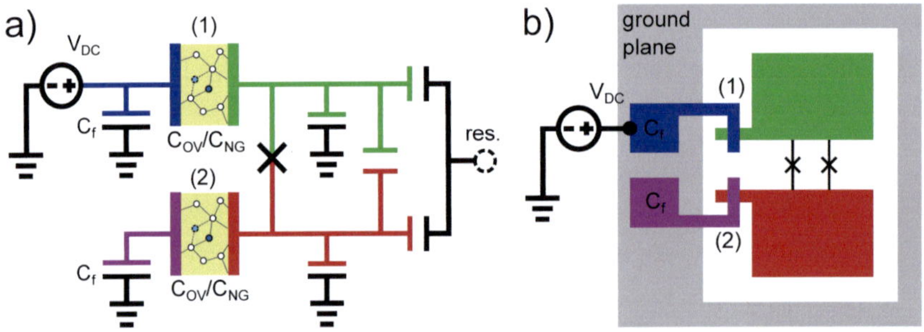

Figure 2.5: **a** Circuit diagram of one out of many TLS-sensor realizations based on a floating transmon qubit and derived from the Xmon-based TLS-sensor illustrated in Fig. 2.4 (b). **b** A corresponding design sketch, where no readout resonator is shown. The transmon electrodes are illustrated by the green and red pads while the top electrodes of sample capacitors are painted in blue and violet. TLS hosted in the sample capacitors (sample-TLS) are tunable by electric fields induced by the dc-voltage V_{DC}. Hereby, the dc potential drop between the qubit electrodes is zero due to the transmon regime, while each sample capacitor experiences a voltage V_i ($i \in \{1,2\}$), where $V_1 + V_2 = V_{DC}$. Since the island electric potential is V_{DC}-dependent, TLS hosted at interfaces (intrinsic TLS) may be tuned by V_{DC}. Thus, the detector realization shown here and another one mentioned in the text do not allow to fully distinguish sample-TLS from intrinsic ones. We believe that this feature, which is naturally present in the Xmon-based detector used in this work, is not achieved with the floating transmon topology without large calibration efforts (see text).

2.3.4 Spectroscopy with applied elastic and electric fields: distinguishing junction-TLS (Theory)

The first and second TLS experiments with ready-made qubits were performed with an Xmon sample chip [Bar+13] from UCSB, Santa Barbara, USA. The chip was installed in the sample box B2 described in Sec. 3.3.3 that allows application of elastic strain (see Sec. 3.3.1) as well as dc-electric fields via global electrodes placed above and underneath the chip as depicted in Fig. 2.6 (a). A voltage V_{tb} was applied between the top and bottom electrode to ascertain that TLS residing on both sides of the qubit film are tuned by the global field. The corresponding circuit schematic is shown in Fig. 2.6 (b), where $C_{t/b}$ is the top/bottom electrode capacitance to the qubit island (green) while $C_{tg/bg}$ is the corresponding capacitance to ground. The dc-wiring in the cryostat fully dominates the latter capacitances so that $C_{tg} \approx C_{bg} \approx 400\,\mathrm{pF}$ (see App. (A) for estimation). Due to this symmetry, the voltage drops across $C_{tg/bg}$ are $V_{tb}/2$ each. Importantly, due to the qubit transmon regime ($E_J/E_C \sim 100$) no dc-voltage drop is induced across the Josephson junction (JJ), so that $V_{tb}/2$ also drops across $C_{t/b}$ (black arrows). Hereby, E_J and E_C are the Josephson and the charging energies defined in Eq. (2.6) and (2.11), respectively.

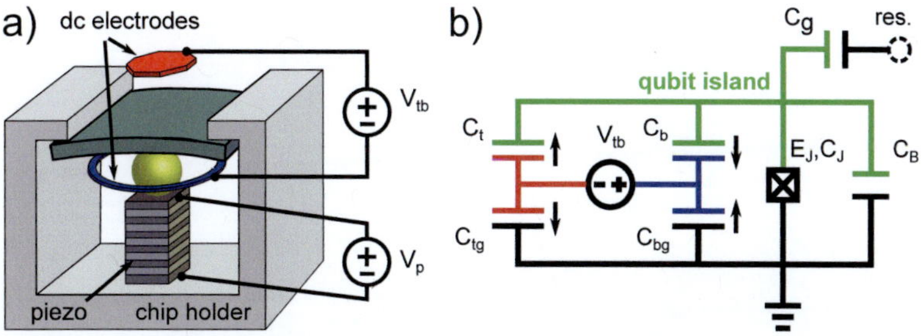

Figure 2.6: **a** Sketch of the sample holder used in the experiment to distinguish TLS hosted in the Josephson junctions (junction-TLS) from TLS located at interfaces of an Xmon qubit. The piezo actuator is used to transfer elastic strain to the sample chip while the global electrodes (red and blue) induce dc-electric fields on the qubit film interfaces. Simulations reported in Sec. 4.2.2 show that the electric field generated by global electrodes and those induced by the qubit are maximum at film edges. Thus, detectable TLS hosted at film interfaces are expected to be tunable via global electrodes as quantitatively explained in the text. **b** Circuit diagram of the Xmon biased by the global electrodes, whose capacitances to the qubit island $C_{t/b}$ are of the order of aF while the qubit total capacitance is $C_\Sigma = C_B + C_J + C_g \sim 90\,\mathrm{fF}$. The electrodes' capacitances to ground are dominated by dc-wiring in the cryostat: $C_{tg} \approx C_{bg} \approx 400\,\mathrm{pF}$. Due to the transmon regime of the qubit, no dc-voltage drop is induced between the electrodes of the qubit Josephson junction (JJ). First, this implies that an equal voltage $V_{tb}/2$ (black arrows) drops in the capacitors $C_{t/b}$ and $C_{tg/bg}$ due to the symmetry of $C_{tg/bg}$. Second, junction-TLS respond to strain only, while TLS located at interfaces are tunable by elastic and electric fields.

The TLS transition frequency ω_{TLS} expressed in terms of applied voltages is

$$\omega_{TLS}(V_p, V_{tb}) = \sqrt{\Delta^2 + (\varepsilon + \gamma_p V_p + \gamma_{tb} V_{tb})^2}, \tag{2.44}$$

where $\hbar\Delta$ is the TLS tunnel energy and $\hbar\varepsilon$ the intrinsic asymmetry energy. γ_p and γ_{tb} are individual TLS sensitivities to elastic and electric fields, respectively. Due to permanent zero dc-voltage across the Josephson junction, TLS hosted inside the JJ tunnel barrier are not tunable by global electric fields ($\gamma_{tb} = 0$). Thus, TLS hosted at qubit interfaces respond to elastic and electric fields, while junction-TLS will respond to elastic fields only.

The setup of Fig. 2.6, where dc-voltage is applied between the top and bottom electrodes, was used for measurements discussed in Sec. 4.2.4. However, it turned out that similar results can be obtained in a simplified setup that employs only one electrode. As reported in Sec. 4.2.2, finite element simulations show that the electric field induced by qubit oscillations is concentrated at film edges and decays as $\sim 1/\sqrt{x}$, where x is the distance to the film edge. Hereby, a TLS having an electric dipole moment of 10 D (maximum imaginable value in solids) and residing at $x > 200\,\mathrm{nm}$ would couple to the qubit by $g/2\pi < 0.05\,\mathrm{MHz}$, which is about the typical detection sensitivity of the UCSB samples. Thus, 200 nm is the maximum distance to the edge where TLS are still detectable. Further, simulations show that at such distance, and on both sides of the film, the dc-electric field created by each separate global electrode is minimum 130 V/m per volt applied (maximum applied voltage is $\pm 100\,\mathrm{V}$). A TLS with a 10 D dipole moment could be thus shifted by maximum 1.3 GHz in its asymmetry frequency $\varepsilon/2\pi$, which is comparable with the maximum observable TLS tunnel frequency $\Delta/2\pi \sim 6.3\,\mathrm{GHz}$ using the UCSB samples. Thus, we can be sure that every detectable TLS is also tunable by any of the two employed global electrodes. Therefore, the simplified setup requires one electrode only as depicted in Fig. 2.7, and allows for distinction of junction-TLS similar to the previous setup.

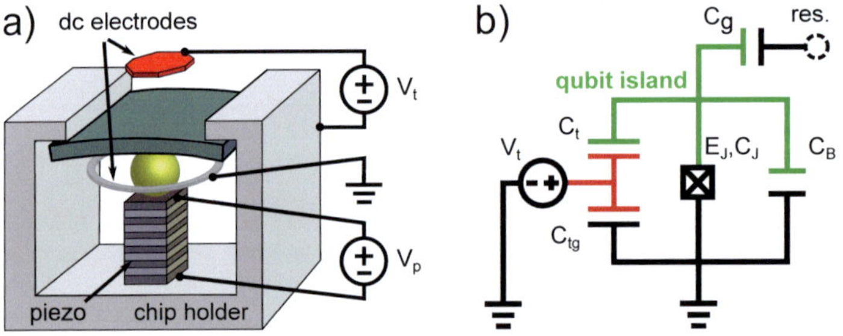

Figure 2.7: **a** Sample holder sketch of the simplified setup that is expected to allow discrimination of TLS hosted in the Josephson junction (junction-TLS) from those residing at other interfaces. The only difference to the previous setup shown in Fig. 2.6 is that the dc-voltage is applied to one global electrode, e.g. the top electrode (red) while the bottom electrode (gray) is grounded. Simulations show that the electric field generated by each separate global electrode is high enough on both sides of the qubit film edge to enable field-tuning of detectable TLS residing at interfaces (see Sec. 4.2.2). **b** Circuit diagram of the Xmon biased by the global electrode, whose capacitance to the qubit island is negligibly small compared to the qubit total capacitance $C_\Sigma = C_\mathrm{B} + C_\mathrm{J} + C_\mathrm{g} \sim 90\,\mathrm{fF}$. The electrode capacitance to ground is about 400 pF dominated by its dc-wiring. Due to the transmon regime, no voltage drop is induced in the Josephson junction (JJ) when applying V_t, so that junction-TLS do not respond to global fields.

The discrimination method shown here fails in an improbable case if the electric dipole moment of a TLS is oriented perfectly perpendicular to the electric field, which implies a small error. Further, this technique does not require any special qubit geometries. It works with ready-made Xmon-type and floating transmon samples but also with qubits in the phase regime ($E_\mathrm{J} \ggg E_\mathrm{C}$). Independently of the $E_\mathrm{J}/E_\mathrm{C}$ ratio, it is applicable to qubits whose topology is that of a superconducting ring interrupted by a Josephson junction, where both JJ electrodes are at same dc-electric potential since they are shorted by the ring.

2.3.5 Deducing TLS positions with two dc-electrodes (Theory)

Here, we discuss an experiment to deduce possible locations of TLS at interfaces close to the film edge of the qubit circuit. For this, we use two separately controlled global dc-electrodes by referencing the voltage sources independently to the qubit chip ground as shown in Fig. 2.8

(a). The corresponding circuit diagram is shown in Fig. 2.8 (b); the piezo actuator is not used in this experiment. By comparing measured and simulated location-dependent TLS coupling strengths to each electrode, we obtain information about the possible position of each TLS. For example, TLS at the substrate-metal interface couple more strongly to the bottom electrode, while TLS at the film-vacuum interface respond more strongly to the top electrode. This is further illustrated by blue and red arrows in Fig. 2.9 (b). The electric field strongly varies along the profile of the film edge, while it is presumably constant along the edge of the qubit island and the enclosing ground plane shown in Fig. 2.9 (a). For this reason, we map the geometry to the two-dimensional problem illustrated in Fig. 2.9 (b) showing the film edge profile, where the colored lines (see legend) indicate the interfaces of interest for this experiment: the substrate-vacuum (SV), substrate-metal (SM), the native oxide (Ox), and the interface of the native oxide and vacuum (OxV). The colored lines also define the spatial axes used to label the TLS location $x^{(i)}$ at a given interface i.

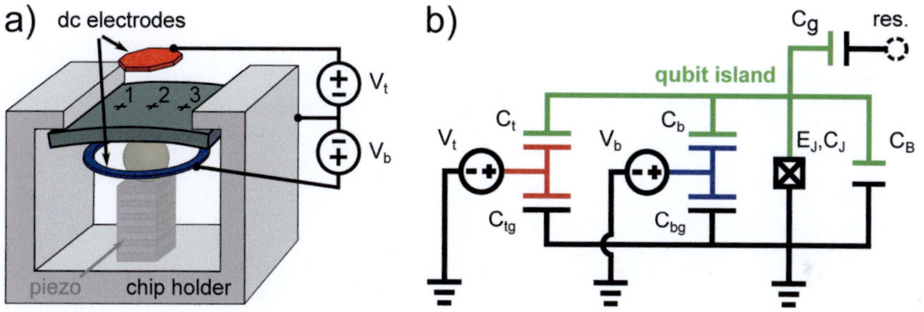

Figure 2.8: a Sketch of the sample holder used in the experiment to resolve TLS positions at Xmon film edges. The piezo actuator is not used here while dc voltages V_t and V_b are separately applied to global electrodes placed above and underneath the sample chip to create dc-electric fields at qubit interfaces. The locations of the three qubits on the chip are indicated by crosses. **b** Circuit diagram of the global electrodes and the qubit connected to a readout resonator. C_t and C_b are the electrode capacitances to the qubit island while C_{tg} and C_{bg} are their capacitances to ground.

In Sec. 4.2.1, the method is presented to simulate the electric fields $\boldsymbol{E}_{t/b}$ at the film edge created by the top and bottom electrode, respectively. Fig. 2.10 (a) contains a plot of $\boldsymbol{E}_{t/b}$ unit vectors along the path indicated in the left inset, for applied negative voltage $V_{t/b}$ to either the top or the bottom electrodes. The electric field orientation at the OxV interface that is not shown here deviates by less than 0.5° from that in the native oxide layer. The black arrows in (a) indicate unit vectors of the qubit-induced field $\boldsymbol{E}_q$. The right inset contains a color plot giving an overview of $|\boldsymbol{E}_t|$ at the film edge for $V_t = -0.5\,\mathrm{V}$ and $V_b = 0\,\mathrm{V}$. In Fig. 2.10 (c), simulated $|\boldsymbol{E}_t(x^{(i)})|$ for same voltages is shown while $|\boldsymbol{E}_b|$ plotted in (b) results from swapped voltages. The TLS position $x^{(i)}$ is defined in the inset of Fig. 2.10 (b) and (c). We drop the index (i) in the following text for better readability.

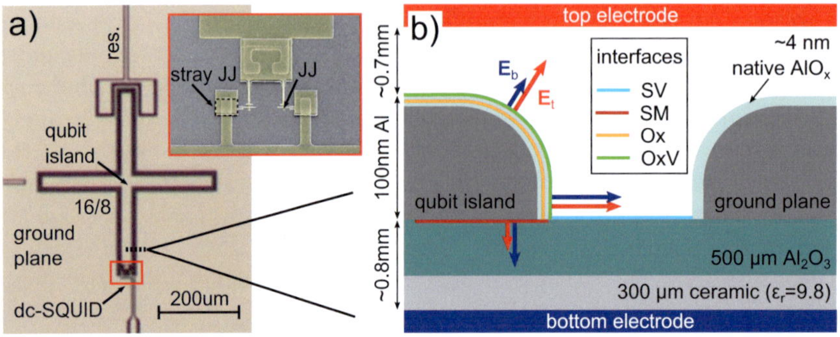

Figure 2.9: a Photograph of a UCSB qubit (Xmon 2) used in this experiment. The inset shows an SEM picture of a qubit's dc-SQUID that consists of two parallel small Josephson junctions (JJs) which are connected in series with large-area stray junctions. The latter are artifacts of the commonly applied tilted-angle deposition technique. **b** Schematic of the film edge profile indicating the interfaces at which TLS positions are investigated: the substrate-vacuum (SV), substrate-metal (SM), the native oxide (Ox), and the interface of the native oxide and vacuum (OxV). The red and blue arrows symbolize the electric field $\boldsymbol{E}_{t/b}$ induced by the global electrodes at three representative locations of the film edge. Depending on the location, a TLS is differently strongly tunable by the top and the bottom electrodes. Thus, from the measured tunability ratio and the simulated ratio $|\boldsymbol{E}_t|/|\boldsymbol{E}_b|$, one can extract possible TLS locations at the film edge profile. The SV-interface analysis is more involved as discussed in the text.

The transition frequency of an interface-TLS has the form

$$\omega_{\rm TLS} = \sqrt{\Delta^2 + \Big(\varepsilon + 2\boldsymbol{d}\big(\boldsymbol{E}_t(x) + \boldsymbol{E}_b(x)\big)/\hbar\Big)^2} \tag{2.45}$$

$$= \sqrt{\Delta^2 + \Big(\varepsilon + 2\big(d_{\parallel t}|\boldsymbol{E}_t(x)| + d_{\parallel b}|\boldsymbol{E}_b(x)|\big)/\hbar\Big)^2}, \tag{2.46}$$

where $\hbar\varepsilon$ is the intrinsic TLS asymmetry energy when no fields are applied. Since we do not apply elastic strain in this experiment, the piezo actuator only adds a constant offset to ε as explained in Sec. 3.3.1. $\boldsymbol{d}$ is the TLS electric dipole moment and $d_{\parallel t/b}$ its component parallel to the corresponding field $\boldsymbol{E}_{t/b}$. In terms of applied dc voltages, we rewrite (2.46) as

$$\omega_{\rm TLS}(V_t, V_b) = \sqrt{\Delta^2 + (\varepsilon + \gamma_t V_t + \gamma_b V_b)^2}, \tag{2.47}$$

where $\gamma_{t/b}$ is the TLS sensitivity to the voltages that results from a fit of the recorded TLS frequency to Eq. (2.47). It comprises the information about TLS location and dipole moment orientation through the identity

$$\hbar\gamma_{t/b}V_{t/b} = 2d_{\parallel t/b}|\boldsymbol{E}_{t/b}(x)|. \tag{2.48}$$

The possible TLS locations $\{x\}$ are solutions of the left and right terms of the following equation:

$$\frac{\gamma_t}{\gamma_b} = \frac{d_{\parallel t}E_t(x)}{d_{\parallel b}E_b(x)} = \frac{\cos(\alpha_d - \alpha_{tb}(x)/2)E_t(x)}{\cos(\alpha_d + \alpha_{tb}(x)/2)E_b(x)} \equiv \zeta(\alpha_d, x)\frac{E_t(x)}{E_b(x)}, \tag{2.49}$$

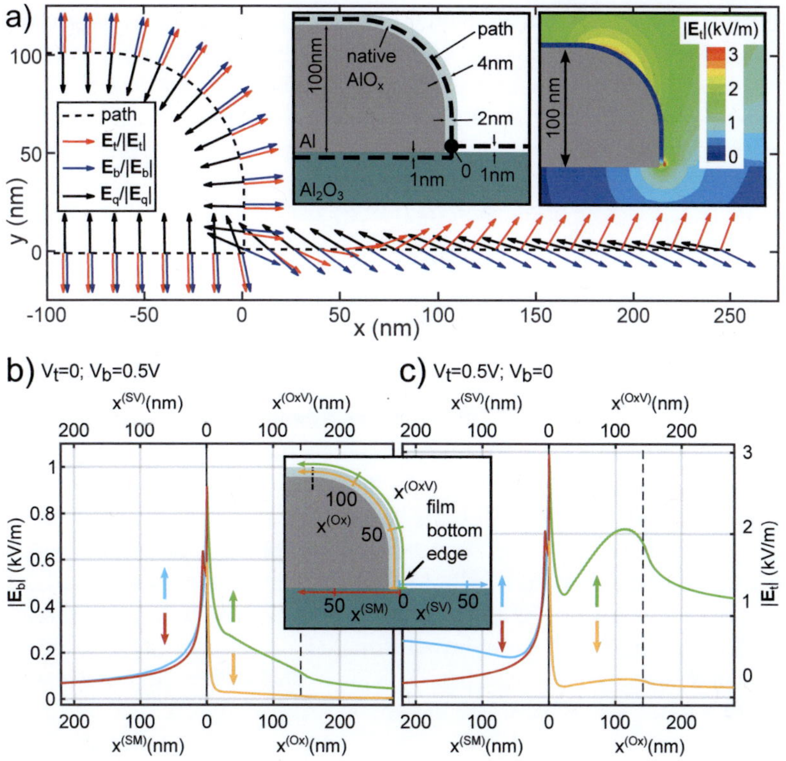

Figure 2.10: a Unit vectors of simulated electric fields $\boldsymbol{E}_{t/b}$ for applied negative voltages either to the top or to the bottom electrodes, as well as the qubit-induced field $\boldsymbol{E}_q$. The dashed line denotes the path at the film edge profile and the substrate-vacuum interface along which the field vectors are plotted (see left inset). It also sets the spatial coordinate x to define the TLS location while $x = 0$ (black dot) denotes the substrate-metal-vacuum edge (also called film bottom edge). The indicated thickness of native aluminum oxide on top of the Al film is assumed to be $\sim 4\,\text{nm}$. The right inset contains a color plot of the simulated absolute electric field $|\boldsymbol{E}_t|$ at the film edge for 0.5 V applied to the top and 0 V to the bottom electrode. **b, c** $|\boldsymbol{E}_t(x^{(i)})|$ and $|\boldsymbol{E}_b(x^{(i)})|$ plotted vs. the TLS location $x^{(i)}$ defined in the inset. The simulation method is presented in Sec. 4.2.1.

where $E_{t/b}(x) \equiv |\boldsymbol{E}_{t/b}(x)|/V_{t/b}$ is the simulated electric field per applied voltage to the top/bottom electrodes. The angle α_d denotes the TLS dipole moment orientation and $\alpha_{tb}(x)$ is the angle between $\boldsymbol{E}_t$ and $\boldsymbol{E}_b$, as defined in Fig. 2.11 (c). From Fig. 2.10 (a), we recognize that at the SM, Ox and OxV interfaces (film interfaces) the fields from both electrodes are parallel. Thus, at film interfaces, the equation (2.49) simplifies to

$$\frac{\gamma_t}{\gamma_b} = \frac{E_t(x)}{E_b(x)}. \tag{2.50}$$

In Fig. 2.11 (a) the simulated ratio E_t/E_b at film interfaces of Xmon 2 is shown. From the measured ratio γ_t/γ_b of an exemplary "TLS 1"(violet line), we derive its possible positions to be $x^{(Ox)} = x^{(OxV)} = 15\,\text{nm}$ at the Ox and OxV interfaces indicated by red dots in Figs. (2.11) (a) and the left panel of (d). From the deduced locations and Eq. (2.48), we calculate the corresponding $d_{\parallel t}$ $(= d_{\parallel b})$ and discard solutions resulting in unreasonably large values $> 10\,\mathrm{D}$. In this representative example, the Ox-location was disregarded (see right panel of Fig. 2.11 d) since the electric field in the native oxide is by roughly $1/\varepsilon_r \approx 0.1$ smaller than at other interfaces. This is explained by a simple calculation of the electric field strength inside and outside (e.g. at the surface to vacuum) of a thin dielectric layer (thickness l) that covers an electrode of a plate capacitor (plate distance $L \gg l$). The remaining solution is degenerate in the TLS dipole orientation α_d ($[0..\pi]$) as it is valid for all possible TLS dipole projections. Generally, at the film interfaces, where $\boldsymbol{E}_t \parallel \boldsymbol{E}_b$, every solution has such degeneracy.

At the SV interface, the global fields are not parallel and solutions of the full equation (2.49) are required. In Fig. 2.11 (b) and (d), derived SV-locations of TLS 1 for a small and exemplary set of α_d values are indicated by black dots. In the analysis of measurement data, we use about a hundred interpolated α_d values in the interval $[0..\pi]$. Similar to solutions at film interfaces, we drop SV-solutions implying $d_{\parallel t/b} > 10\,\mathrm{D}$. The remaining possible SV-locations of this example are $x^{(\mathrm{SV})} \in [5..40]\,\mathrm{nm}$ with corresponding $\alpha_d \in [0.05..0.45]\,\pi$. TLS 1 is thus located either at $x^{(\mathrm{OxV})} = 15\,\mathrm{nm}$ with a probability $p^{(\mathrm{OxV})} = \pi/(\pi + 0.4\pi) = 0.71$, or at $x^{(\mathrm{SV})} \in [5..40]\,\mathrm{nm}$ with a probability $p^{(\mathrm{SV})} = 0.4\pi/(\pi + 0.4\pi) = 0.29$. Here, the probabilities are estimated from the size of the solution space spanned by allowed α_d which at film interfaces is trivially 2π per solution x. At the SV interface, it is given by the number of x solutions after truncation (black dots in Fig. 2.11 (d), right) times the infinitesimal angle element ($\pi/10$ in this simplified example). In our data analysis, all interface-TLS are treated individually in this way to obtain statistics on TLS locations. The average likelihood for an analyzed TLS to reside at a given interface i is then defined as

$$P^{(i)} = \sum_{\{\mathrm{TLS}\}} p^{(i)} \Big/ \sum_{\{\mathrm{TLS}\}}, \tag{2.51}$$

where the summation goes over all investigated TLS.

This method is based on the assumption that the applied electric field is constant along the qubit edges, which is valid for the qubit Xmon 2 due to its central position. For Xmons 1 and 3, the E_t/E_b ratio varies by a factor of 0.5 up to 1.8 along the qubit edge due to their eccentric positions relative to the global electrodes. This method is thus applicable to Xmon 2 only, as reported in Sec. 4.2.5. A simple and important upgrade of the setup would be large electrodes comparable to the chip size, which would ensure field homogeneity all over the chip.

As indicated in Fig. 2.9 (b), the top corner of the film edge is rounded due to dry etching during the sample fabrication. Our simulations show that the ratio $E_t(x)/E_b(x)$ does not significantly change when varying the curvature radius from the assumed value of 70 nm to 30 nm. Thus, estimation errors of the top edge curvature do not influence our analysis on TLS location. For the same reason, we argue that the ratio $E_t(x)/E_b(x)$ does not change at the edge profile of shadow-evaporated films used to form the qubit Josephson junction in the UCSB samples examined in this work.

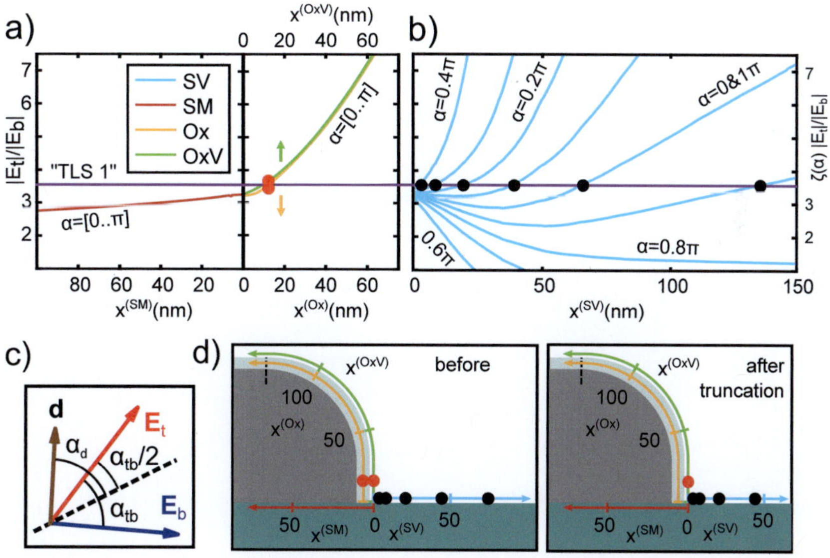

Figure 2.11: Possible locations of an exemplary "TLS 1" detected in Xmon 2. **a** Simulated ratio $E_t(x^{(i)})/E_b(x^{(i)})$ of the dc-electric fields created by global top/bottom electrodes, along the SM, Ox and OxV interfaces (film interfaces) introduced in Fig. 2.9 (b). Here, $x^{(i)}$ is the spatial axis at each interface introduced in (d), and defines the TLS location. $E_{t/b} \equiv |\boldsymbol{E}_{t/b}|/V_{t/b}$ is the simulated electric field per applied voltage to the top/bottom electrodes. The intersections (red dots) of the measured tunability ratio γ_t/γ_b (violet line) with $E_t(x)/E_b(x)$ indicate possible TLS locations that are solutions of the equation (2.50). E.g., TLS 1 may be located at $x^{(Ox)} = 15\,\text{nm}$ and $x^{(OxV)} = 15\,\text{nm}$. Since $\boldsymbol{E}_t$ is parallel to $\boldsymbol{E}_b$ at film interfaces, each solution is degenerate in the TLS dipole moment orientation α_d ($[0..\pi]$) defined in (c). **b** At the SV interface, each TLS location is associated with one α_d value as indicated in Eq. (2.49). The solution set depicted by black dots spans the positions $x^{(SV)} \in [5..140]\,\text{nm}$ for angles $\alpha_d \in [0..0.4; 0.9..1]\pi$. For better readability, a small and exemplary set of α_d is evaluated here, while for the analysis we use about a hundred interpolated α_d values in the interval $[0..\pi]$. **c** Definition of the angles between the directions of the applied fields and the TLS dipole moment. **d** From Eq. (2.48) and the deduced location, the electric dipole moment component $d_{\parallel t/b}$ parallel to $\boldsymbol{E}_{t/b}$ is deduced. Each solution requiring $d_{\parallel t/b} > 10\,\text{D}$ is discarded. The left and right panels indicate the set of solutions (red and black dots) before and after such truncation, respectively. TLS 1 is located either at $x^{(OxV)} = 15\,\text{nm}$ with a probability $p^{(OxV)} = \pi/(\pi + 0.4\pi) = 0.71$, or at $x^{(SV)} \in [5..40]\,\text{nm}$ with a probability $p^{(SV)} = 0.5\pi/(\pi + 0.4\pi) = 0.29$. Here, the probabilities are estimated from the size of the solution space spanned by allowed α_d. This analysis is repeated individually for each TLS to obtain statistics.

3 Experimental Details

3.1 The experimental setup

3.1.1 Cryogenic setup

In order to use a micro-fabricated circuit as a superconducting quantum device, one has to operate it at low temperatures using a cryogenic refrigerator. It is crucially important to cool the quantum circuits to even much lower temperatures than their superconducting transition temperature in order to suppress thermal noise and thermal excitations of the superconducting state, both perturbing the coherence of the devices. In this work, all the samples have been measured in the Oxford Kelvinox 100 dilution refrigerator (DR) at a temperature of 30 mK. Here, the DR working principle is roughly explained.

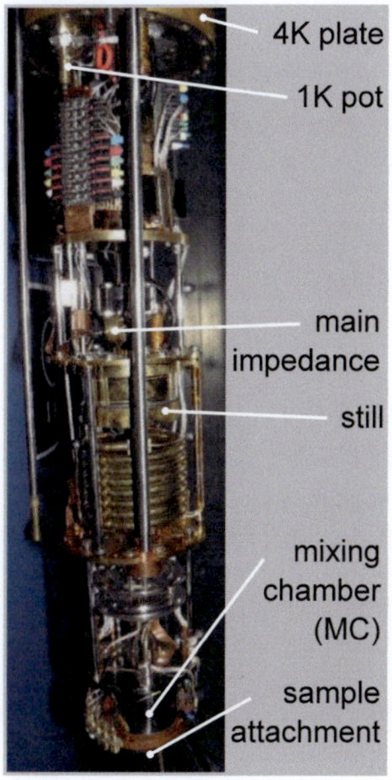

Figure 3.1: A photograph of the Oxford Kelvinox 100 insert.

The cryogenic medium of a DR is a ^{3}He/^{4}He isotope mixture enclosed in a circulation loop. In the mixing chamber (MC) which defines the coldest spot in the refrigerator, at temperatures below 870 mK, two separate phases of the mixture coexist. The so-called rich phase is practically pure liquid ^{3}He swimming on top of the diluted phase consisting of superfluid ^{4}He with some few percent of ^{3}He. From the MC, the diluted phase is transported upwards into the still that is kept at low pressure at around 0.1 mbar, see Fig. 3.1. Here, ^{3}He evaporates out of the superfluid ^{4}He that is at rest. Once evacuated out of the cryostat, the ^{3}He gas passes the circulating pump at room temperature. Then, it is filtered via cold traps at 77 K and at 4 K, after which it condenses due to thermal contact with the 1K pot filled with liquid ^{4}He. The 1K pot is a pumped chamber kept at about 1.7 to 1.9 K due to the lowered boiling point of the liquid ^{4}He that is fed from the liquid ^{4}He bath through an adjustable impedance. After condensation, ^{3}He passes the main impedance and then the heat exchangers on its way towards the MC. In the exchangers, it thermalizes with the cold ^{3}He that travels upwards to the still. In the MC, ^{3}He diffuses into the diluted phase while the osmotic pressure difference is maintained by pumping out the ^{3}He from the still. Mixing of ^{3}He and the diluted phase necessitates heat absorption from the environment which translates into the cooling power of the MC. The piece of the circulation loop from the 1K pot to the still is placed in the inner vacuum chamber (IVC) that is dipped into the liquid ^{4}He bath for thermal isolation. Without any cable wiring, the MC reaches a base temperature of about 15 mK. Once all cryogenic RF and dc cables are installed, the minimum temperature increases to about 20 mK due to the heat transport via the cables.

3.1.2 Cryogenic wiring and measurement setup

The application of microwave pulses for quantum device manipulation requires coaxial wiring inside the fridge. Although cold RF and dc wiring is a well established cryogenic technique, one still needs care and patience when handling it. Between the cryostat cool-downs, it is recommended to check the tightness of screws on clamps that thermalize the cables at every temperature stage. A loose clamp can lead to increased thermal noise at the sample or to a thermal short of the mixing chamber (MC) to the warmer stages. Also, the cables that reach from the 4K plate down to the MC should have a low thermal conductivity to ensure a proper thermal isolation between the stages. For RF applications, one typically uses stainless steel cables for a good trade-off between attenuation ($\sim$ 10 dB/m@8 GHz) and thermal isolation. An expensive alternative are superconducting cables that have a very low attenuation ($<$ 1 dB/m@8 GHz) and an ideal thermal isolation. For the wiring within a temperature stage, such as the MC, high conductance cables from copper or from silver-plated metals are typically used. For dc cables, one usually uses superconductor or constantan thin wires that are bound in a bunch and are thermalized at every stage with a stycast glue.

The first task of this PhD project was to upgrade the cryogenic wiring to enable dispersive readout of the qubit quantum state. This necessitates a probe pulse to be sent down to the cold sample and be transferred back after scattering on the readout resonator. The according installed input and output coaxial lines are shown in Fig. 3.2 b). The input line is heavily attenuated at every temperature stage to thermalize the inner conductor and to reduce the thermal noise coming from the room temperature tools. At the MC, the band-pass filter (BP) removes the low- and high-frequency noise of the input signal. In the sample box, the probe signal is fed into the transmission line patterned on the sample microchip, where it interacts with the notch type readout resonator and then exits the sample box. After passing another BP and two isolators, the signal is passed through a superconducting coaxial cable to the high electron mobility transistor(HEMT)-based amplifier at 4 K. Hereby, the BP serves as a dc-break protecting the HEMT while the isolators serve for thermalization of the thermal noise from outside. Back at ambient temperature, the output signal is amplified by two room-temperature amplifiers before being processed by the readout apparatus. In addition, infrared filters are installed at every port of the sample box, which were custom-made by Oleksandr Lukashenko. They absorb stray light from outside or from the dissipative attenuators and filters in the line [Cór+11]. The second line shown in Fig. 3.2 b) is differently attenuated and filtered, and used to tune the qubit frequency by calibrated flux pulses.

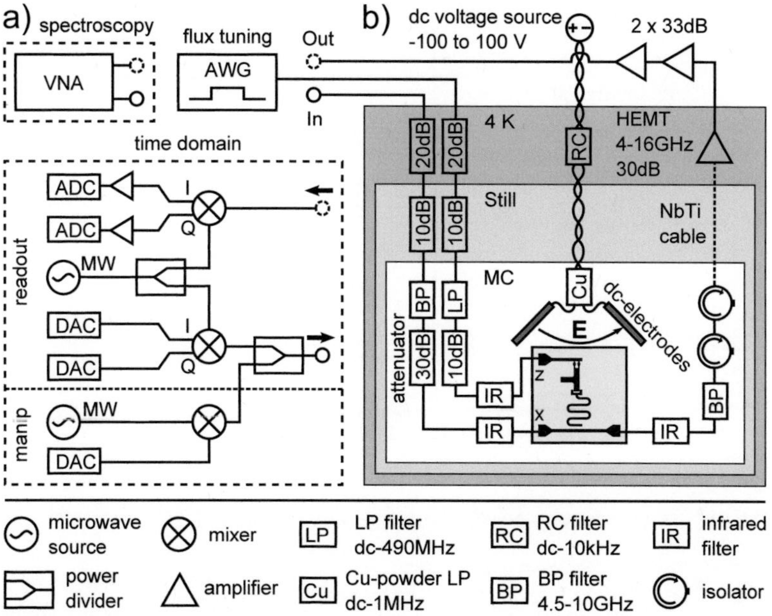

Figure 3.2: a Schematic of the standard spectroscopy and time domain setup used to characterize the readout resonators and to detect the qubit quantum state, respectively. The qubit frequency is tunable via calibrated flux pulses generated with an AWG. **b** Drawing of the microwave wiring at the cryostat temperature stages. The blueish rectangle symbolizes the microchip containing the T-shaped split-junction transmon qubit coupled to a transmission line via a readout resonator. Typically, we measured samples containing three independent qubits. The global dc-electrodes and the piezo actuator (not shown) are controlled via twisted-pair dc-lines which are filtered as illustrated here.

The global dc-electrodes illustrated in Fig. 3.2 b), and further explained in Sec. 3.3.3, are controlled by a voltage source via twisted-pair dc-lines. The latter lead through an RC and a copper powder low-pass filters [LU08] whose cutoff frequencies are $\sim 10\,\mathrm{kHz}$ and $\sim 1\,\mathrm{MHz}$, respectively. The piezo actuator introduced in Sec. 3.3.1 is similarly addressed via dc-lines with an RC filter that has a $\sim 1\,\mathrm{Hz}$ cutoff frequency.

Since in our experiments we exploit the qubit as a reading head to detect individual Two-Level-Systems (TLS), it is sufficient to apply standard single-qubit readout schemes to gain information about TLS properties. Before starting the qubit measurements, we first characterize the resonators using a vector network analyzer (VNA) device. The two dashed boxes in Fig. 3.2 a) show the spectroscopy setup using a VNA and the homodyne readout setup used in time domain measurements. To manipulate the qubit quantum state, we apply pulses to the readout line using a power combiner. All the apparatus and microwave parts used here are listed in Tab. 3.1.

Device	Model
Microwave source 1	Agilent E82576
Microwave source 2	Anritsu MG37022A
Arbitrary wave generator	Telektronix AWG 5914B 1.2 GS/s
HEMT cold RF amplifier	Low Noise Factory AB LNF-LNC4-16A 4-16 GHz
room temp. RF amplifier	Mini-Circuits VA-183-S
Cryogenic circulator	QuinStar QCY-060400CM00 magnetic shielding, 4-8 GHz

Table 3.1: Components of the time domain setup for pulsed microwave measurements as shown in Fig. 3.2.

3.2 Micro-fabrication techniques

3.2.1 Optical lithography and alignment with the E-Beam writer

The experiments on TLS-sensors derived from a transmon qubit were performed on samples fabricated at KIT. Xmon-type qubits were fabricated in a 2D coplanar architecture using apparatus listed in Tab. 3.2. Standard fabrication techniques were used combining a first step optical lithography (OL) and a second step electron beam lithography (EBL) to pattern the main structures and the Josephson junctions (JJs), respectively. Typically and unfortunately, this process requires in total three lithography steps since alignment marks (EB marks) for the electron-beam (EB) writer are necessary. In a non-industrial clean-room, every additional lithography step diminishes the yield and increases the time consumption, which was a strong motivation to facilitate this process. Here, a trick is presented to circumvent the extra lithography step required for the EB-marks.

Device	Model
electron beam writer (see Tab. 3.3)	JEOL JBX-5500ZD 50 keV acceleration voltage
mask aligner	Carl Suess MA6, Xe 500 W lamp, wave lengths: 240, 365, 405 nm
inductively coupled plasma (ICP) device	Oxford Plasma Technology Plasmalab 100 ICP180 variable substrate temp. 273 – 343 K
shadow evaporation device	PLASSYS MEB550s
wafer dicing device	DISCO DAD3350

Table 3.2: Apparatus used to fabricate TLS-sensor samples at KIT.

The qubit samples were fabricated as follows (see App. C for details): All structures except the JJs were dry etched in a 50 nm thick Al ground plane on a sapphire substrate using an ArCl (Sec. 5.1) inductively coupled plasma (ICP) and S1804 optical resist as a mask. The JJs were patterned with a standard Niemeyer-Dolan double angle evaporation technique [NK76; Dol77] using a suspended bridge developed into a resist stack (A-4 PMMA on top of EL-13 copolymer). Such a fabrication succession implies two disadvantages, namely the need for an additional lithography step to form EB marks and the herewith implied alignment errors. EB marks need a good material contrast to be detectable by the EB writer (JEOL, 50 keV accel. voltage), which is not given by 50 nm Al on sapphire. For this reason they have to be deposited in an extra step using another metal or aluminum thicker than 100 nm. This requires alignment of the EB marks and the structures patterned in the ground plane using a manual mask aligner, which implies an inaccuracy of up to $\pm 2\,\mu m$ that dominates the faults of the EB writer. This circumstance was solved in the following procedure: A same photo-resist mask is used to first pattern the main structures into the ground plane via plasma etching, and to successively form the EB alignment crosses via lift-off. Prior the deposition of the EB mark material (here 70 nm Nb, see Sec. 5.4.1), the wafer is covered with a protecting wafer leaving small margins for the alignment crosses. This technique is illustrated in Fig. 3.3, it not only saves one lithography step, but it also skips the manual optical alignment since the EB marks are formed with the same mask as the ground plane structures such that the only inaccuracy of the JJ alignment is due to the EB writer ($\pm 1\,\mu m$).

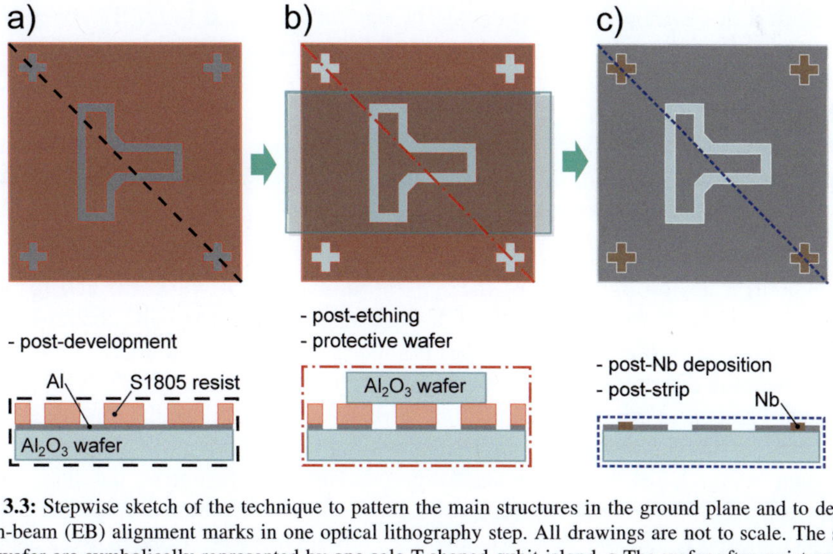

Figure 3.3: Stepwise sketch of the technique to pattern the main structures in the ground plane and to deposit the electron-beam (EB) alignment marks in one optical lithography step. All drawings are not to scale. The six chips on the wafer are symbolically represented by one sole T-shaped qubit island. **a** The wafer after resist application and development, the cross section is shown in the inset below. **b** After etching with an ArCl plasma (Sec. 5.1) the wafer is covered with a protecting wafer leaving a margin with the mask of the EB alignment marks. **c** After Nb deposition, removal of the protecting wafer and stripping of the chip with N-Eethyl-2-pyrrolidon (1 h at 90 °C). The EB-marks are self-aligned to the ground plane structures and ready for the subsequent EB lithography to form the Josephson junctions (see Sec. 3.2.2).

The application of the protecting wafer does not harm the main structures that are protected by the resist profile. Manufacturing a specific metallic frame that grips sample and protection wafers would facilitate and speed up this process. There are other possibilities to circumvent the extra lithography step for EB marks, e.g. by using an Al ground plane thicker than 100 nm or inverting the order of OL and EBL. The first possibility was canceled since we decided to use a 50 nm thick Al film to avoid interruptions of the subsequently deposited JJ films at the first layer edges. The second possibility would have implied additional contaminations due to the lift-off process used after the OL, and again a high misalignment risk.

3.2.2 Fabrication of the Josephson junctions

As described in the previous section, first the main structures are dry etched in the 50 nm Al film on a C-plane oriented sapphire wafer and then Josephson junctions (JJs) are formed after an electron beam lithography (EBL). For good galvanic contact, the native oxide of the existing Al film is removed using an in-situ Argon plasma milling recipe [Gr17] prior deposition of the Al layers that define the JJs. The JJs are formed using the Niemeyer-Dolan technique [NK76; Dol77] with a resist mask consisting of ~350 nm A-4 PMMA on top of ~850 nm EL-13 copolymer. In the PLASSYS shadow evaporator, 30 nm Al at an angle of 24° and 60 nm Al at –24° are deposited. Between the deposition steps, static oxidation takes place for typically 4.25 min at 15 mBar oxygen pressure, which forms the JJ tunnel barrier due to thermal oxidation of Al. With a typical Junction size of 350 nm×350 nm, the normal resistance of a dc-SQUID from two equal JJs was $Rn = 4.25\pm0.1\,\mathrm{k\Omega}$. By comparing the theoretically predicted maximum qubit frequency f_{01} shown in Eq. (2.14) with the measured values ~6.5 GHz of four qubits, the critical temperature of Al was calculated to be $T_c = 1.14\,\mathrm{K}$ from the standard BCS relation $\Delta_s(T = 0\,\mathrm{K}) = 1.764\mathrm{k_B T_c}$ and the Ambegaokar-Baratoff relation $I_c = \pi\Delta_s(T)/2eR_n$, where $\Delta_s(T)$ is the temperature-dependent superconducting gap. Using the last relation, the SQUID critical current was estimated to about $I_c = 65\,\mathrm{nA}$ and the critical current density $J_c = 265\,\mathrm{nA}/\mu\mathrm{m}^2$.

In Fig. 3.4 (a) a design overview of the Xmon qubit is given while (b) and (c) are the magnifications of the dc-SQUID and the JJ designs, respectively. The table contains the layer colors associated with the lithography step number and the layer number. The green area No. 3k depicts the dry etched Al film from the first OL step while the other layers define the EBL mask used to form the dc-SQUID. The charge densities deposited by the EB writer and are listed in Tab. 3.3 along with the corresponding layer number. An SEM record of the JJs is shown in Fig. 3.4 d), where the narrow double fingers indicated by the blue arrow are artifacts from a side-project that do not interfere with the experiments in this work. The two notches on the qubit island and the two teeth on the ground plane are made to avoid accidental interruptions of the JJ films by the underlying Al film edges, which is a problem inherent to the double angle technique. One sees small "mouse bites" at the edge of the underlying film that are reported to stem from sub-film oxidation [Ric+16] that is accelerated by JJ annealing but also forced by Chlorine etching residuals [Hes82; DDH87; Pro+99; RL95] that react with air humidity. In Sec. (5.2), more details are reported concerning such void formation in Al films on sapphire substrates. To trace out the weekly EBL fluctuations influencing the JJ area and resulting in a JJ resistance deviation of up to 1 kΩ, JJs were fabricated on the same day on multiple identically prepared wafers. During the pumping of the PLASSYS chamber (~2 h) to prepare the second wafer for junction formation, the first (previously processed) wafer was stripped, the resistance of its dc-SQUIDs was measured at room temperature, and the target oxidation exposure E_{t} for the following wafers was deduced using an empirical exponential law $J_c \propto E^{-\alpha}$ [KMM95]:

$$E_{\mathrm{t}} = E_{\mathrm{m}} \left(\frac{R_{\mathrm{t}}}{R_{\mathrm{m}}} \right)^{1/\alpha}, \tag{3.1}$$

where $E = tP$ is the exposure defined as the product of the oxidation time t and the static oxygen pressure P. The exposure of the first wafer is E_{m} while its measured mean SQUID resistance is R_{m}. From the target SQUID resistance R_{t}, the target exposure E_{t} for the following wafers is then deduced. The empirical exponent is reported to be about $\alpha \approx 0.4$ in the exposure range of 10 to 10^4 mBar · s [KMM95]. From fits of Eq. (3.1) to the SQUID resistance of four identically fabricated samples oxidized for 170 to 330 sec at 15 mBar (2500 to 5000 mBar · s), α was deduced to about 4.9 in this system.

layer No	scale factor
1	2.475
4	2.2
100	0.5
101	0.75

Table 3.3: The mask used to form the Josephson junctions and the overlap junctions was exposed in the JEOL 50 keV EB writer with a 2 nA current, 8 nm spot size and 10 nm spot distance. The global charge density was 435 μC/cm^2 while the four layers defining the exposure pattern had individual charge density scale factors listed above. The layer 1 and 4 define areas of fully developed double resist while the layers 100 and 101 define the suspended resist structures. Hereby, layer 1 is used to form mostly narrow trenches with a well-defined resist edge while layer 4 is used to define large resist-free areas with a rather uneven edge. The layer 100 is used as a spacer between the layer 1 and 101 to smoothen the doses gradient and prevent the overexposure of the suspended resist edges.

While developing the JJ design, about 370 test JJs and dc-SQUIDs were fabricated on five wafers. During further design optimization 180 test- and 180 Xmon-SQUIDs were fabricated on ten wafers each containing six chips with the full qubit design. After annealing at 200 °C for 5 min, the test dc-SQUIDs showed an average resistance increase by 4±8 %. On two of those wafers overlap junctions (OVs) (Sec. 2.3.2) were fabricated that required backing of the wafers at 200 °C for 2 × 5 min under an EL-13 copolymer resist. Their dc-SQUID resistance showed a decrease by 2.5±1.3 %. The resistance increase after the first annealing is attributed to resist residuals that remain after development [KVM07; NH07]. The resistance decrease after the second annealing however is linked to the adsorption of hydrates ($-OH$) from the resist by the tunnel barrier [Pop+12]. To avoid JJ contamination with residuals, an isotropically cleaning OAr plasma was applied in the PLASSYS (see recipe in the appendix, B.1) before the Ar milling and the subsequent JJ deposition. Regarding the post-annealing resistance decrease by

50 % for JJs fabricated on uncleaned substrates [KVM07; NH07], this recipe seems effective and comparable to a more aggressive cleaning with a Reactive Ion Etching oxygen plasma that was reported to fully cancel the residual-related resistance increase [Pop+12].

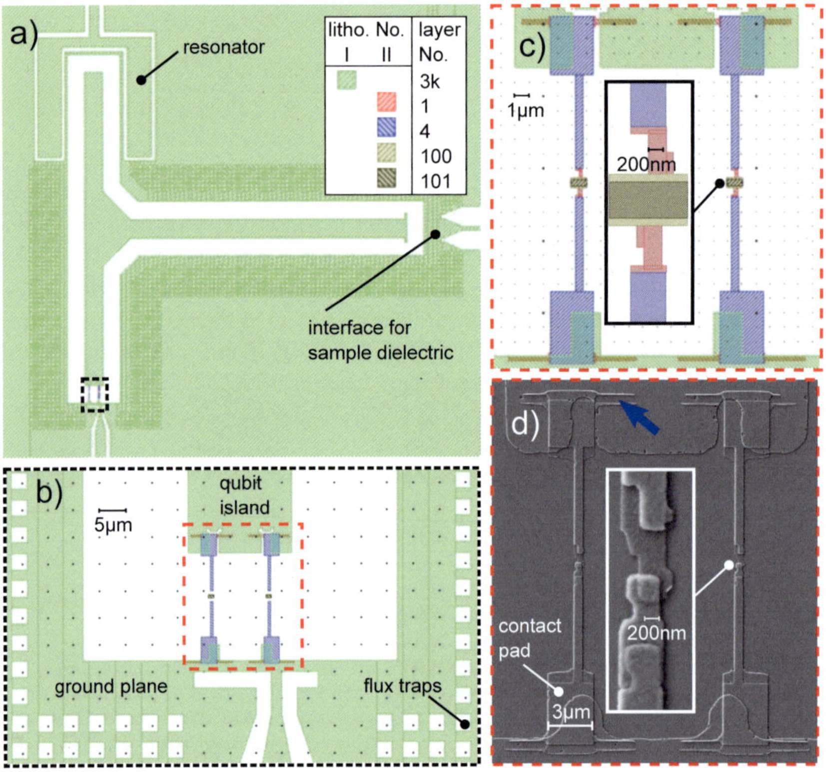

Figure 3.4: **a** Design overview of the fabricated Xmon. The layer 3k denotes the qubit island and the ground plane formed in the first optical lithography step. The other layers define the mask patterned in the double resist in an electron beam lithography process used to form the Josephson junctions (JJs). The corresponding charge densities are listed in Tab. 3.3. **b** Magnification of the dc-SQUID. **c** Design of the JJs. **d** SEM record of a dc-SQUID containing two equal junctions. The "mouse bites" in the ground plane and qubit film edges are due to chlorine residuals from dry etching, as further explained in 5.2. The double-fingers (blue arrow) are artifacts from another side-project, and do not have a special role nor influence here.

The EBL resolution depends on the substrate and on metallic structures formed in previous fabrication steps. Finding the right doses pattern was time consuming due to proximity effects caused by the doses pattern but also by the metallic electrodes under the resist stack. Connected to the latter issue, I learned that the pads of test SQUIDs should emulate the surrounding of the qubit SQUIDs. This is helpful when quickly characterizing the oxidation result on the test SQUIDs placed on the same wafer with the qubits. This is further reported in Sec. 5.3.

3.2.3 Fabrication of the overlap contacts

The overlap contacts (OVs) consist of a capacitor made from two Al electrodes separated by 50 nm of amorphous insulating AlO_x. OVs were made on wafer L3 with areas of 300 nm × 2.1 µm and on wafer XI1 with areas of 300 nm × 300 nm, having capacitances of 1.1 fF and 0.16 fF, respectively. The bottom electrodes of large OVs were made in an optical lithography (OL) step, while for small OVs, they were made in the same eBeam lithography (EBL) step as the qubit Josephson junctions. Sample dielectric and top electrodes of both OV types were formed in the PLASSYS using an A-4 PMMA/EL-13 copolymer mask patterned in an EBL step: After removing the native oxide of the bottom electrode with an Ar milling process [Gr17] (see recipe in the appendix, B.1), the AlO_x layer was formed during a perpendicular deposition of Al at a rate of 0.2 nm/s in an oxygen atmosphere (chamber pressure of 3×10^{-4} mBar, oxygen flow of 5 sccm). The dielectric was in-situ covered by perpendicularly deposited 100 nm of Al that formed the top electrode.

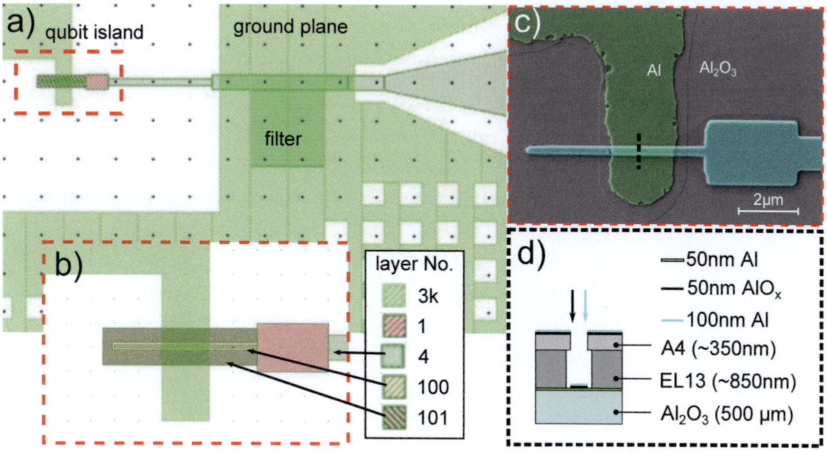

Figure 3.5: a Design overview of the large overlap junction (OV) used on wafer L3. It constitutes a plate capacitor containing a 50 nm thick insulating AlO_x sample dielectric. **b** Zoomed-in view of the OV. Layer 3k depicts the design of the main structures patterned in the ground plane from the first optical lithography step. The protrusion of the qubit island is the OV bottom electrode. The OV dielectric and the top electrode are deposited in a lift off process. The corresponding mask is written in an electron beam lithography process using the charge pattern defined by the layers 1 to 101, whose doses are listed in Tab. 3.3. The large plate capacitor (250 fF) serves as a low pass filter and as an effective grounding of the OV top electrode. This enables the tuning of the sample TLS with a dc electrical field applied to the OV top electrode while preserving the qubit coherence. **c** SEM picture of the large OV. The green shaded layer is the OV bottom electrode while the turquoise layer is the sample dielectric covered with the OV top electrode. **d** OV cross-section sketch before lift-off of the mask, where the arrows denote the deposition angle of OV materials.

In Fig. 3.5 (a) and (b) an overview and a zoomed-in view of the large OV design on wafer L3 are shown. The layer No. 3k depicts the structure formed in the OL step, namely the ground plane and the qubit island with a protrusion serving as the OV bottom electrode. The other layers denote the charge pattern deposited with the eBeam writer to define the mask used to form the OV dielectric and the top electrode. An SEM image of an OV is shown in Fig. 3.5 c), where the green shaded layer is the bottom electrode and the turquoise colored layer is the OV dielectric covered with the top electrode. Fig. 3.5 d) contains a sketched OV cross-section before lift-off of the mask, while the arrows denote the deposition direction of the AlO_x and Al. The relatively uneven film edges of the bottom electrode visible in (c) are possibly related to an unexpected change of the ICP processing chamber environment that occurred after a refurbishment of its turbo pump. For the following samples, the recipe was re-optimized, see Sec. 5.1. While the AlO_x layer poses the sample dielectric in the OV, it also serves as a spacer between the ground plane and the transmission line leading from the bonding pads to the OV top electrode. The herewith formed large capacitor of $C_f \sim 250\,\mathrm{fF}$, serves as a filter shown in Fig. 3.5 (a) and introduced in Sec. (2.3.2). It suppresses RF noise from the transmission line and serves as a DC-break in the qubit circuit. Further, the usage of such plate capacitor enables both ground plane sides along the transmission line to be kept at same potential that suppresses parasitic slot line resonances. The filter creates a clean microwave environment for the qubit while allowing for electric field tuning of TLS contained in the OV dielectric by applying a dc voltage to the OV top electrode.

In Fig. 3.6 (a) to (c), the respective design overview, magnified view and an SEM picture of a small OV used on wafer XI1 are presented. To save one lithography step, the bottom electrode is formed in the same second lithography step as the qubit junctions shown in Fig. (3.4) (b). It consists of an Al layer (blue) galvanically coupled to the qubit film and covered by a second Al film (red) separated by a tunnel barrier of the same thickness as that in the qubit JJs. Such a large-area stray JJ is inherent to the double angle JJ formation technique. In Fig. 3.6 (d), the formation of an asymmetric undercut in the double-resist [Lec+11] is illustrated that is used to cut the second (red) layer as shown in the inset (i). This prevents the stray JJ shown in (ii) from crossing the OV top electrode. The OV dielectric and top electrodes are deposited after a third step (EBL) as shown in (iii), similar to the large OV described above. The small islands at the end of the OV electrode strips have two purposes. First, they add some charge proximity during the EB writing to cleanly form the narrow trenches that define the electrode strips. Second, they facilitate the developer flow through the trenches to optimize the development rate of the exposed resist.

3.2.4 Fabrication of the nano-gap contacts

The so-called nano gap (NG) is a coplanar realization of overlap junctions (OVs), whose advantage is cleaner interfaces. The OVs are cross junctions made in two lithography steps, thus before depositing the sample dielectric and the top electrode, an Argon plasma cleaning step is required to remove the native Al oxide on the bottom electrode. This may introduce contaminations due to unwanted back-deposition of sputtered resist. Both NG electrodes as well as the sample dielectric are deposited in-situ and successively, without any electrode cleaning, which implies cleaner interfaces. The fabrication of a transmon qubit equipped with an NG required four lithography steps. The qubits were formed in the first two steps, namely an optical lithography (Ol) and an E-beam lithography (EBL) step as usual. In the following third OL step, the ground plane was covered with an insulating 50 nm thick AlO_x spacer (see Fig. 3.7 (a)) to separate it from the NG electrodes that were formed in the last EBL step. The NG fabrication is based on a bridge free double angle Josephson junction (JJ) formation technique [Lec+11] illustrated in Fig. 3.7 (b) and the insets (i) to (iii). In contrast to the original JJ formation process, the electrodes do not overlap, but they miss each other forming a constant gap of ~50 nm. Also the static oxidation between the electrode depositions is skipped. There was no extra dielectric deposited on top of the NG shown here. In such case, the NG can be used to detect TLS hosted in the native Al oxide formed on the electrodes as well as TLS formed due to adsorbates on the oxide surface. Due to time limitations the NG samples were not measured in this work.

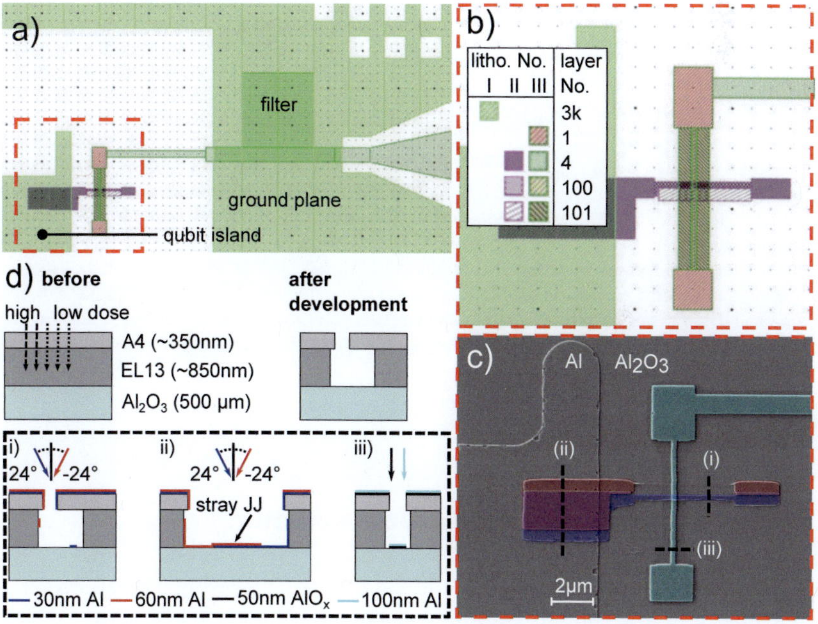

Figure 3.6: Small overlap junction (OV) fabricated on wafer XI1 that constitutes a plate capacitor containing a sample dielectric (here 50 nm of insulating Al_2O_x) **a** Design overview of the OV. **b** Magnified OV view. The table associates the eight patterns to the lithography step number and to a layer number. The layer No. 3k depicts the design of the qubit island and the ground plane patterned in the first optical lithography step. The other layers constitute the charge pattern for the two following electron beam lithography (EBL) steps. **c** SEM picture of an OV, whose bottom electrode is formed in the same EBL step as the Josephson junctions (JJs). **c** Illustration of the asymmetric undercut [Lec+11] formation used to cut the second layer (red) as shown in the inset (i) to prevent the stray JJ shown in (ii) from crossing the OV top electrode. The OV dielectric and its top electrode were deposited in a third step (EBL) as shown in (iii).

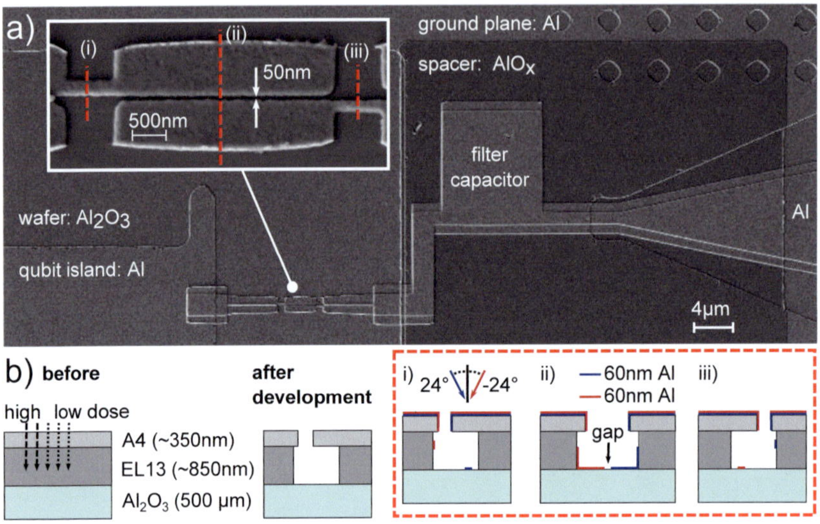

Figure 3.7: **a** SEM image of a nano gap contact (NG) that is a coplanar capacitor with a relatively small gap of ~50 nm. The sample dielectric can be deposited in-situ on top of the NG, which guarantees a clean dielectric in contrast to the overlap junctions formed in two separate lithography steps and presented in the previous section. When no sample dielectric is deposited like in this case, the NG serves to detect TLS in the native Al oxide on the electrodes, as well as TLS formed by adsorbates on the oxide surface. **b** The NG fabrication was inspired by the bridge free technique to form Josephson junctions [Lec+11] by exploiting an asymmetric undercut of the resist mask. To form an NG, however the electrodes do not touch each other forming a gap, and no oxidation between the electrode depositions is used. The corresponding deposition sequences are shown in the insets (i) to (iii).

3.3 The sample housing

3.3.1 Strain-tuning of TLS using a piezo-actuator

Tuning TLS by direct control of the strain in the sample chip using a piezo actuator is one central technique employed in this work. It was initiated by prof. G. Weiss and developed at KIT by T. Peichl, G. Grabovskij et al. [Pei12; Gra14]. Here, its working principle is briefly described. The 6 mm × 6 mm × 0.5 mm sapphire chip is fixed in a metallic holder while a stack piezo actuator [Pie] is docked to its backside to elastically bend it (see sketch in Fig. 2.6 a). An applied voltage from -50 to 80 V controls the piezo elongation that is about 13 nm/V at 30 mK. This factor is not affected by the chip stiffness that is 25 times smaller than that of the piezo. From simplified geometrical calculations and from finite element simulations [Pei12; Gra14], the surface elongation coefficient $\Delta l/l$ is determined to be 3 to $7 \cdot 10^{-7}$ /V. Hereby, l is the length of a line on the chip top side hosting the sample film and Δl is its elongation per Volt. Since the thermal contraction of the piezo actuator is much smaller than that of the copper or brass piezo holder, the chip is pre-strained during the cool down, which adds a constant offset to the intrinsic TLS asymmetry ε_0 (see, e.g., Eq. (2.41)). More importantly, this allows one to strain-tune the TLS not only by piezo elongation but also by its contraction.

3.3.2 Sample box B1: shielding aspects of the qubit

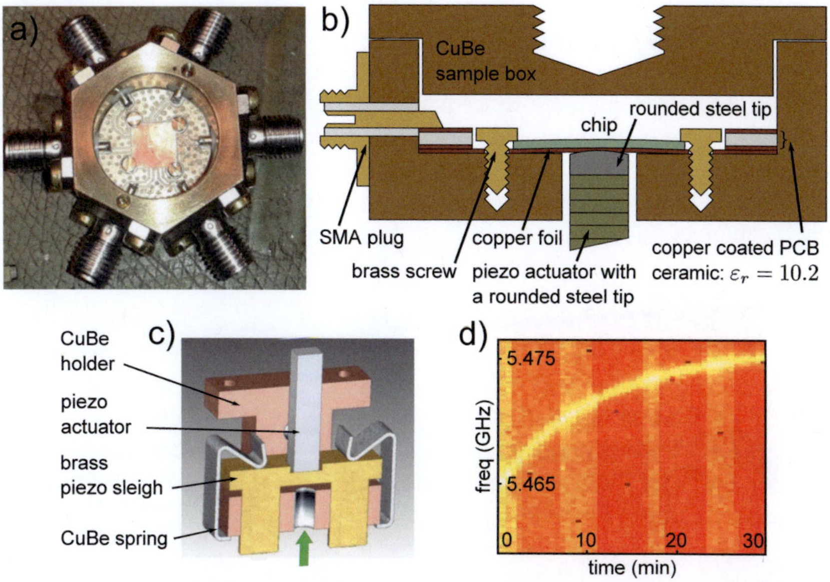

Figure 3.8: **a** Photograph of the opened first sample box version B1. **b** Cross-section sketch of B1. The chip is tightened by four brass mini screws and elastically bent by the piezo actuator that pushes from beneath. **c** Cross-section of the piezo holder used with B1. The piezo is docked to the chip before the cool-down via a screw (green arrow) underneath the piezo sleigh. Hereby, the voltage between the piezo electrodes is observed to detect the beginning actuator contraction. **d** In B1, the qubit suffered from direct heating by the piezo actuator. The plot shows a typical relaxation of the qubit resonance frequency to the initial value after a fast drop caused by switching the voltage applied to the piezo actuator from −20 V to 20 V. After replacing the steel tip (see b) of the piezo by a zircon sphere, as well as improving the piezo thermalization (see Fig. 3.9 a), no such qubit resonance creeps were observed anymore.

During the upgrade of the cryostat, wiring for dispersive qubit readout, i.a., a sample housing was developed that offered some advantages in comparison to the original sample holder used in the works of T. Peichl, G. Grabovskij et al. [Pei12; Gra14]. In Fig. 3.8 a), a photograph of the first housing version (B1) machined from Copper Beryllium (CuBe) is shown. Its first advantage is the metallic cavity that encloses the chip and protects it from environmental noise, radiation losses and electromagnetic crosstalk from the piezo. Second, the chip is mounted from the top, held by four mini screws as shown in the sketch (b). Moreover, the piezo holder shown in (c) has become more compact. Similar to the original sample holder, a copper foil separates the chip back side and the piezo actuator. However, the zircon sphere that was supposed to protect the piezo from shear deformations is replaced by a rounded steel tip. We observed in this setup that after each voltage change applied to the piezo, the qubit resonance frequency dropped by several MHz and relaxed back on a time scale of minutes as shown in Fig. 3.8 (d). We explain this by heat transfer from the piezo to the chip via the steel tip. Presumably, this problem occurred or was at least reinforced by the fact that B1 was connected at its top lid to the cryostat cold finger, while the piezo holder from CuBe was attached to the bottom of the box. Such a constellation and the relatively low thermal conductivity of CuBe may lead to heat accumulation at the piezo actuator.

To reduce the earth's magnetic field around the qubit, all sample housing versions were placed at the bottom of a closed, 30 cm high, 18 cm wide, and 1 mm thick Cryoperm shield. However, we observed in B1 sudden qubit resonance jumps by a few MHz when applying small voltage steps to the piezo actuator. We explain this by vortices trapped in the qubit film due to residual magnetic field inside the box that are moved by the magnetic field induced by the piezo wiring. This would lead to a small change of the magnetic field close to the qubit dc SQUID, and to qubit detuning.

3.3.3 Sample box B2: global dc-electrodes

In the second sample box version (B2), the steel tip of the actuator is replaced by a zircon sphere serving for thermal isolation of the chip (see Fig. 3.9 b and c). Further, the fixation order is changed as shown in Fig. 3.9 (a) and the piezo holder and sleigh are made of copper (see Fig. 3.8 b) for better thermalization of the piezo. After this upgrade, we didn't observe any qubit frequency creeps from piezo heating.

B2 is made from aluminum for better shielding of the sample from magnetic fields. In this housing, qubit frequency jumps appeared rarely in comparison to observations with B1 and only after applying large voltage steps beyond 20 V to the piezo actuator. This behavior is possibly due to the 4.2 mm wide holes in the box drilled for the SMA microwave plugs shown in Fig. 3.9 (a). Although the holes trap the earth's magnetic flux once the box becomes superconducting, their large diameter allows for local flux density fluctuations within the hole, which still can be enough to move vortices trapped in the qubit film. A practical solution turned out to be repetitive sweeping of the piezo voltage within its maximum working range over night, which seems to anneal vortices in the qubit film by stray magnetic fields from the piezo wiring. Another explanation is that vortices are moved by elastic shock waves generated at chip edges that rub against the chip fixation.

Two global electrodes are integrated in the housing B2, enabling one to tune TLS hosted at qubit film interfaces via electric dc-fields. The top electrode was glued to the housing lid while the annular bottom electrode was patterned on the backside of the printed circuit board (see Fig. 3.9 b and c). From dc-field tunability characteristics, it is possible to identify TLS hosted in qubit's Josephson junctions as reported in sections (2.3.4) and (4.2.4). Comparison of respective coupling strengths of a TLS to the top and bottom electrodes allows for derivation of possible TLS locations at the qubit film edges (see Sec. 2.3.5 and (4.2.5)).

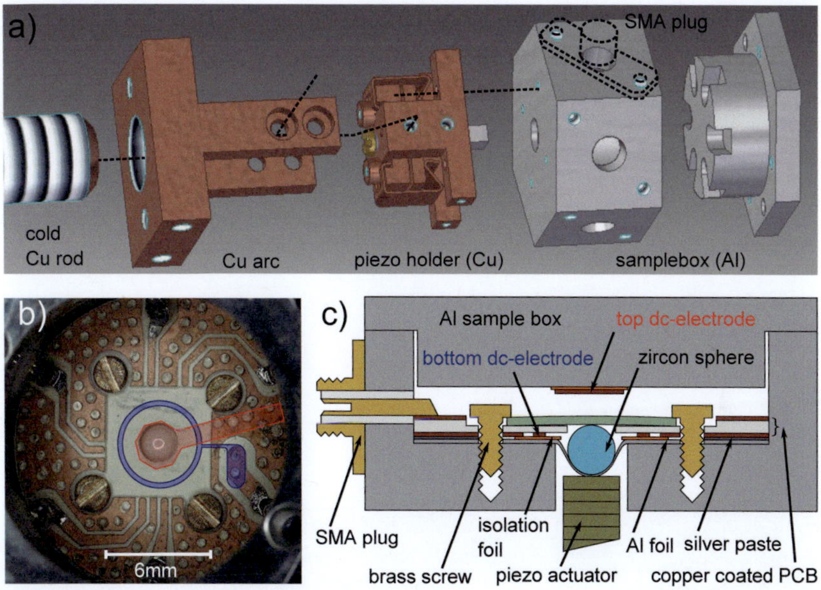

Figure 3.9: **a** Exploded view of the second sample box version B2 and its fixation that is also used for the third box version B3. In contrast to version B1, the piezo holder was made of copper and attached directly to the mixing chamber cold finger, which improved dissipation of the heat produced in the piezo actuator. **b** Photograph of the opened B2 with an illustration of the global electrodes used in the experiments on finding TLS locations at the qubit film edges (see Sec. 2.3.5 and (4.2.5)). The top electrode (red) was glued to the housing lid while the bottom electrode (blue) was patterned into the printed circuit board back side. **c** Drawing of the B2 cross-section. The zircon sphere serves for thermal isolation between the piezo actuator and the sample chip while the Al foil and the box serve for electromagnetic protection.

3.3.4 Sample box B3: non-magnetic parts

The third sample housing version B3 is completely made from non-magnetic materials and the holes used for RF-connectors have become smaller (diameter of 1.2 mm) to reduce the amount of trapped vortices in the qubit film. This did not improve the behavior of sudden qubit frequency jumps observed in the box version B2. However, B3 has some other advantages. It allows for direct soldering of the coaxial cables to the microwave lines on the printed circuit board (PCB) in the box (see Fig. 3.10 a), which avoids the impedance mismatch and stray magnetism from RF plugs. B3 has a cavity underneath the sample chip that promises less crosstalk between the RF lines on the chip [Wen+11]. Moreover, it allows for a better alignment of the zircon sphere due to a ceramic frame positioned at the bottom of the cavity. The box resonance modes of B3 were simulated using the eigen-mode solver of the finite element simulation tool ANSYS HFSS. The lowest box mode is 14.6 GHz, well above the the typical readout resonator frequencies of maximum 7.75 GHz. The experiments on qubit-based TLS-sensors reported in sections (4.1) were performed using B3.

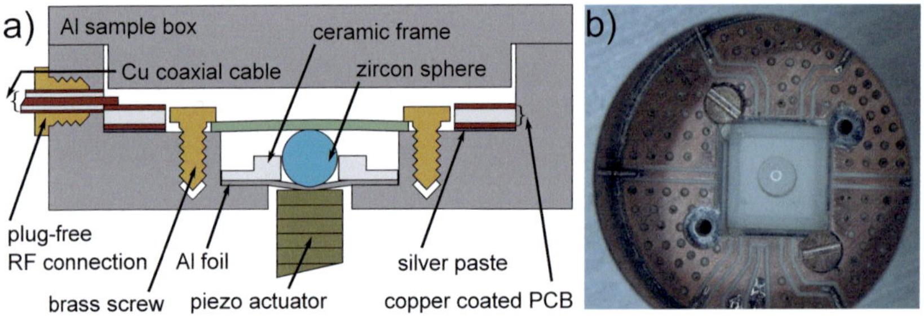

Figure 3.10: The latest version of the sample box (B3) used in the experiments with qubit-based TLS-sensors reported in Sec. 4.1. **a** Drawing of the B3 cross-section that was completely made from non-magnetic parts. The plug-free connection serves for an impedance-matched soldering of the coaxial cables onto the printed circuit board inside the box. The cavity underneath the sample chip is used to reduce the crosstalk of the on-chip RF lines [Wen+11] while the ceramic frame aligns the Zircon sphere. **b** Photograph of an empty and opened sample box with two mini screws.

4 Results: strain- and E-field spectroscopy of TLS in Xmons

4.1 TLS-sensor: spectroscopy with local dc-electrodes

The two-level system (TLS) sensor introduced in Sec. 2.3.2, constitutes an Xmon qubit shunted with a small capacitor containing a sample dielectric of interest. Hereby, the qubit induces an oscillating voltage across this capacitor, allowing for detection of individual TLS hosted inside of the sample. In Sec. 4.1.1, qubit parameters of the fabricated TLS-sensors are presented. Section 4.1.2 contains results of the sample dielectrics' loss tangent estimation while Sec. 4.1.3 includes reports on detection of individual TLS using a qubit-based TLS-sensor, and discussion on the detected TLS properties.

4.1.1 Qubit parameters

The first version TLS-sensor was used to detect TLS residing inside of insulating AlO_x of a plate capacitor that is also called the overlap contact (OV). To detect an individual TLS with a qubit in the time domain, the TLS-qubit coupling strength g should be comparable or larger than the energy relaxation rate of the qubit. The detection and coherent manipulation of single TLS worked out well in the previous experiments [Gra+12; Bil+17] at KIT using a phase qubit with an energy relaxation time of 100 ns only. Thus, for the first TLS-sensor version we aimed at an energy relaxation time $T_{1,\mathrm{Q}} \sim 1\,\mu\mathrm{s}$, corresponding to a minimum required g of $1/\mu\mathrm{s}$. Martinis et al. [Mar+05] observed for TLS hosted inside AlO_x to have a minimum dipole moment component $d_\parallel$ parallel to the qubit field of about $\sim 0.5\,\mathrm{D} = 0.1\,\mathrm{e\AA}$. To obtain the maximum value for the thickness t_{OV} of the AlO_x sample dielectric, this value and the designed qubit parameters (resonance frequency $f_{01} \approx 6.5\,\mathrm{GHz}$, qubit total capacitance $C_{\mathrm{qub}} \approx 110\,\mathrm{fF}$), are entered into Eq. (2.34) to obtain 70 nm. We chose $t_{\mathrm{OV}} = 50\,\mathrm{nm}$ to have the potential of detecting smaller $d_\parallel$, knowing that the sensitivity of the qubit used in [Mar+05] was limited by the short energy relaxation time of $\sim 10\,\mathrm{ns}$. Expecting the energy relaxation rate of a qubit without OVs to be about $\Gamma_0 = 0.1/\mu\mathrm{s}$ and using equations (2.39) and (2.40), the maximum area of a single OV is estimated to about $0.65\,(\mu\mathrm{m})^2$. In this calculation, the loss tangent 1.6×10^{-3} of a thermally grown AlO_x tunnel barrier of a Josephson junction (JJ) was used [Mar+05]. For the first TLS-sensor patterned on the wafer labeled L3, the OVs constituted cross junctions of area $A = 300\,\mathrm{nm} \times 2.1\,\mu\mathrm{m}$, whose fabrication is described in Sec. 3.2.3.

The wafer L3 contains six chips, from which the chip L3.3 was bonded in the sample box B3 (see Sec. 3.3.4) and is shown in Fig. 4.1 (a). One chip contains three identically designed qubits equipped with a varying number of OVs. Every qubit is capacitively coupled to one end of a $\lambda/2$ readout resonator, whose other end is capacitively coupled to the transmission line. Hereby, the resonator-qubit coupling strength g_q was derived from the simulated mutual capacitance of the resonator claw to the qubit island. The mean $g_q/2\pi$ of 64 ± 2MHz was deduced from the measured dispersive shift $-g_q^2/\Delta$ that is the resonance frequency shift between the resonators' few-photon state and its bright state [Koc+07; Ree+10]. Here, $\Delta/2\pi = f_{01} - f_{res}$ is the difference between the qubit and the resonator resonance frequencies. Fig. 4.1 (b) contains a photograph of the T-shaped transmon qubit, whose design is strongly oriented on the Xmon samples [Bar+13] used in the experiments reported in Sec. 2.3.5. The insets of Fig. 4.1 (b) show SEM pictures of the dc-SQUID Josephson junction on the left and an OV on the right. The chips L3.3 and L3.5 were measured in the same cryogenic run. In Tab. 4.1, important qubit parameters are listed. The one-photon resonance frequency f_{01} and the qubit anharmonicity $\alpha/2\pi = f_{12} - f_{01}$ are detected using continuous wave spectroscopy while $\alpha/2\pi$ is twice the distance between the one-photon-driven 01 and the two-photon-driven 02 resonance peaks. The total qubit capacitance C_{qub} is calculated from the relation $\hbar\alpha \simeq -E_C = -e^2/2C_{qub}$ valid in the transmon regime defined by the ratio $E_J/E_C \approx 100$. Hereby, E_J is the maximum Josephson energy of the qubit dc-SQUID and E_C the qubit charging energy. $T_{1,Q}$ results from a fit to the decaying population of the qubit excited state recorded with the time domain protocol shown in Fig. 2.3 (b).

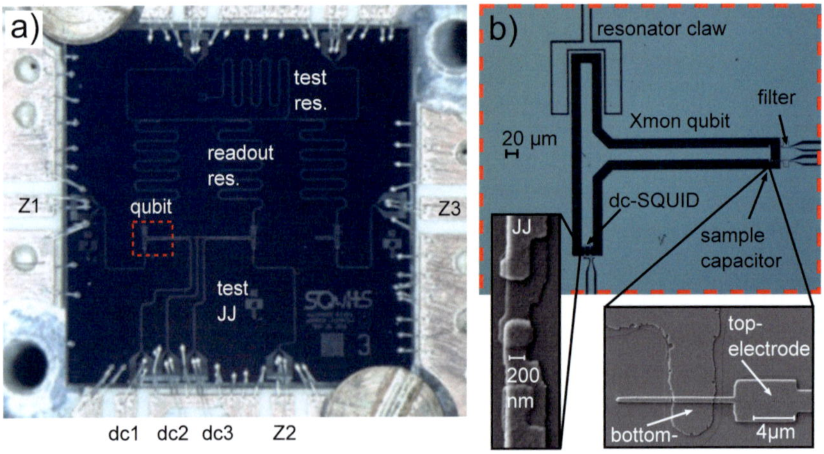

Figure 4.1: **a** Photograph of the sample chip L3.3 tightened in the non-magnetic sample box B3 (Sec. 3.3.4). The chip contains two TLS-sensors and one reference qubit (right) having one, two and no overlap junctions (OVs), respectively. The RF ports Z1-3 are used to tune each individual qubit resonance frequency by magnetic flux pulses. TLS hosted in the AlO_x sample dielectric of the OVs are tunable by electric fields induced by applying voltage to the OV top electrodes (dc channels dc1-3). **b** Photograph of a T-shaped Transmon qubit whose architecture is oriented on the Xmon samples [Bar+13]. The insets show a Josephson junction (left) and an OV (right) whose fabrication processes are described in the respective sections (3.2.2) and (3.2.3). The filter is used to protect the qubit from noise as explained in Sec. 2.3.2.

4.1.2 Deducing the loss tangent of the sample dielectric

The AlO_x sample dielectric was formed by thermal Al deposition (0.1 nm/s) in an oxygen atmosphere (oxygen flow of 5 sccm and chamber pressure of 3×10^{-4}mBar) using a PLASSYS double angle deposition device. The participation ratio $p_{OV}(n)$ of n sample capacitors (called overlap junction (OV)) is deduced from Eq. (2.40) using the qubit total capacitance values C_{qub} listed in Tab. 4.1 while the single OV capacitance is deduced from Eq. (2.35) to be 1.1 fF assuming the AlO_x relative permittivity $\varepsilon_r = 10$ [DOL14; Mar+05]. To deduce the quality factor Q_{OV} of the sample capacitor, these results are entered in Eq. (2.39), where Γ_0 is interpreted as the unperturbed energy relaxation rate when the qubit has no OVs. Accordingly, for chips 3 and 5, we assign the respective values of the reference qubit's $T_{1,Q}^{-1}$ to Γ_0. The resulting averaged AlO_x loss tangent (inverse Q_{OV}) is $(1.7 \pm 0.2) \times 10^{-3}$ that is in good agreement with other experiments: the thermally grown thin AlO_x in a Josephson junction [Mar+05] fabricated in two optical lithography steps showed a loss tangent of 1.6×10^{-3}. The 5 nm thick AlO_x created via plasma oxidation and embedded in $100 - 600\,\mu m^2$ large capacitors of microwave resonators showed $(1.6 \pm 0.2) \times 10^{-3}$ [DOL14]. In a recent experiment at KIT with coplanar waveguide resonators terminated to ground by plate capacitors which were filled with 50 nm thick AlO_x grown via anodic oxidation, losses of about $(0.8 \pm 0.5) \times 10^{-3}$ [Bre+17] were found. In another experiment, where CPW resonators were fully covered with ~500 nm thick AlO_x similarly fabricated as in this work, the loss tangent is reported to be $(1.15 \pm 0.15) \times 10^{-3}$[Pap+11].

qub. No.	Nbr. OV	R_n (kΩ)	f_{01} (GHz)	$\alpha/2\pi$ (MHz)	C_{qub} (fF)	$T_{1,Q}$ (μs)	f_{res} (GHz)	Q_L (10^3)
3.1	1	4.33	6.40	-178	109	1.4	7.23	6.6
3.2	2	4.31	6.47	-172	113	1.0	7.36	11.0
3.3	0	4.35	6.40	-200	97	5.1	7.53	7.9
5.1	1	4.09	6.58	-166	117	0.4	7.22	7.8
5.2	2	4.05	6.60	-174	111	0.55	7.38	4.6
5.3	0	4.37	6.40	-184	105	3.1	7.59	5.6

Table 4.1: Parameters of the first version TLS-sensors (No. *.1 and *.2) and the reference qubits (No. *.3) fabricated and characterized in this work. The columns contain the measured normal resistance R_n of the qubit dc-SQUID, the detected maximum qubit resonance frequency f_{01}, the measured transmon anharmonicity α and the deduced total capacitance C_{qub}. The last two columns show the readout resonator resonance frequencies f_{res} and the loaded quality factors Q_L. The internal quality factor is estimated from fits to quadratures of the test resonators to be $\sim 3 \times 10^5$ and $\sim 6 \times 10^5$ in the few- and many-photon state, respectively. The mean coupling frequency $g_q/2\pi$ between the qubit and the readout resonator is (64 ± 2) MHz. As described in Sec. 4.1.2, from the number of the overlap junctions (Nbr. OV), the respective energy relaxation time $T_{1,Q}$ of the sensor and that of the reference qubit (no OVs), the loss tangent of the AlO_x sample dielectric was calculated to be about 1.7×10^{-3}, comparable to results in similar systems [Mar+05; Pap+11; DOL14; Bre+17].

When designing the first TLS-sensors fabricated on the wafer L3, the same calculation was used to deduce the required minimum OV area to obtain $T_{1,Q} \sim 1\,\mu s$. However, the resulting $T_{1,Q}$ was very low due to the overestimation of the unperturbed energy relaxation time of 10 μs. Further, in contrast to the destructive readout used in the previous works [Gra+12;

Bil+17] that lasted few nanoseconds only, the dispersive readout includes the resonator ring up time $2Q_L/2\pi f_{res} \approx 200$ ns and about 300 ns sampling time in our setup, which is comparable to the sensor $T_{1,Q}$. Thus, our first estimation of a minimum $T_{1,Q} \sim 1$ µs being sufficient for the dispersive readout was too optimistic and led to a bad signal to noise ratio and thus a rather poor TLS visibility. Further, the method shown in section 2.3.1 to determine the minimum detectable coupling strength g breaks down for such short $T_{1,Q}$. Nevertheless, from the measurement data shown in the next paragraph, we can state that the sensor sensitivity is not worse than the minimum detected $g/2\pi$ of ~ 0.9 MHz. To improve the sensor coherence, we reduced the OV area to 300 nm × 300 nm in TLS-sensors on the following sample, expecting $T_{1,Q}$ of an OV-equipped qubit to exceed 3 µs. Due to time limitations, these samples haven't been tested in this work, while the fabrication method is reported in ec. 3.2.3.

4.1.3 TLS detection under elastic and electric fields

Here, the first detection of individual TLS hosted in a sample dielectric (sample-TLS) using a qubit-based TLS-sensor is presented. The color plot in Fig. 4.2 (a) has been recorded with the qubit sample L3.5.2 using the swap-spectroscopy protocol shown in Fig. 2.3 (a) while applying elastic strain to the sample chip via a piezo actuator as described in Sec. 3.3.1. We call such a measurement piezo swap-spectroscopy (PSS). The x-axis shows the voltage V_p applied to the piezo while the color encodes the averaged qubit excited state population after the swap pulse. Similarly, the plot in Fig. 4.2 (b) was recorded while sweeping the dc voltage V_{DC} applied to the top electrode of the first OV, while the top electrode of the second OV was grounded. We call such measurements voltage swap-spectroscopy (VSS). Every bright line indicates a TLS resonance frequency. The theoretical fits to equation (2.41) are shown in Fig. 4.2 (c) and (d), where V_{DC} and V_p are set to zero, respectively.

In contrast to the PSS, there are many horizontal lines in the VSS, which are TLS that do not respond to the applied V_{DC}. Such TLS are hosted in the tunnel barrier of the qubit JJ or in the second OV, where no voltage drop is induced since the qubit island is at the same potential as the ground plane when applying V_{DC}. This is due to the qubit being in the transmon regime far away from the Coulomb blockade, which means that Cooper pairs can easily tunnel through the JJ and compensate the constant potential drop between the qubit island and the ground plane. One can argue that the horizontal lines in the VSS stem from surface TLS far away from the OV. Simulation results (see Sec. 4.2.1) show that surface TLS having a typical dipole moment of 2.5 D [Mar+05] have a maximum qubit-coupling strength of $g/2\pi \approx 0.36$ MHz when they reside at the film edge where the field is focused. This is beyond the sensitivity of this TLS-sensor. The weakly tilted lines, however may indicate surface TLS hosted within a $\sim$100 nm distance from the OV electrodes. The single line that is horizontal in both plots is probably a junction TLS, whose elastic dipole moment is perpendicular to the strain vector that we observe very rarely ($\sim$1/GHz per PSS per cool down).

The TLS numbered from 1 to 4 indicated in Fig. 4.2 (d) are not visible in the PSS for the three following possible reasons. First, the transition frequency of TLS 1 is probably

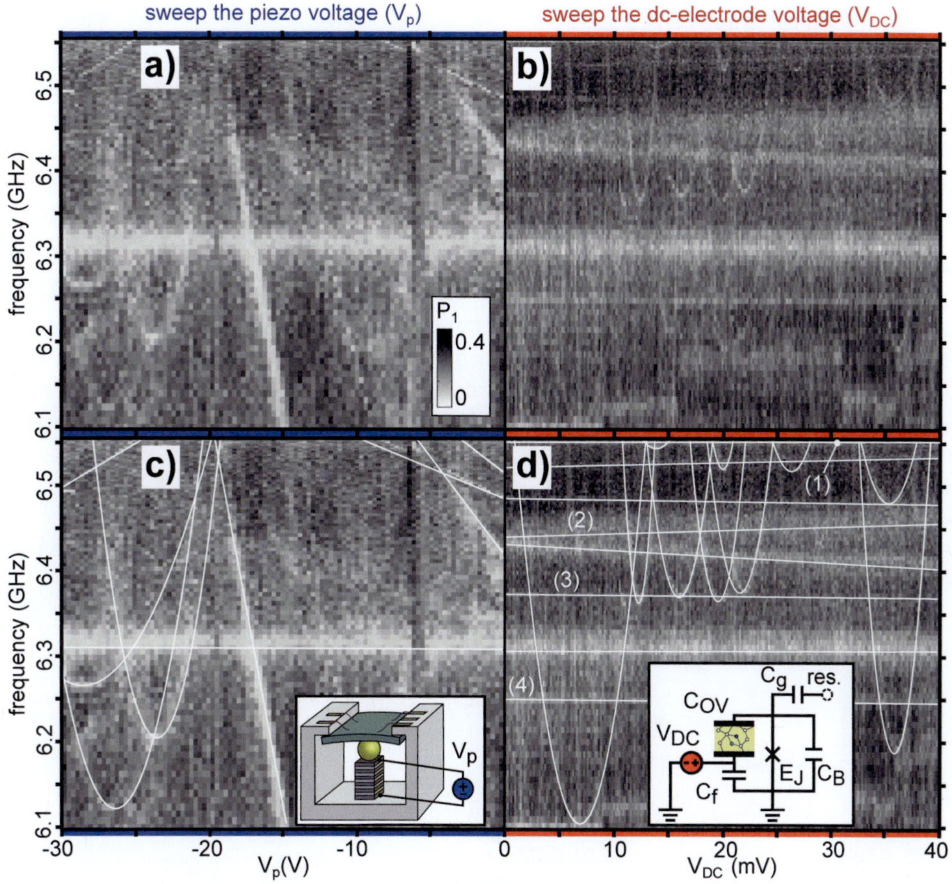

Figure 4.2: TLS distribution recorded with the first version TLS-sensor L3 derived from an Xmon qubit. The sample dielectric consists of 50 nm thick amorphous AlO_x placed in a plate capacitor C_{OV}. **a** TLS distribution under elastic strain of the chip controlled by a voltage V_p applied to the piezo actuator (see inset of (c)). The TLS detection protocol is shown in Fig. 2.3 (a) while the color encodes the qubit excitation population P_1 in dependence of its resonance frequency. Every bright line indicates a resonant TLS that absorbs the qubit energy and leads to a P_1 drop. **b** TLS distribution under application of a dc voltage V_{DC} to the top electrode of the sample capacitor (see right inset of Fig. 4.1 b). The circuit diagram is shown in the inset of (d). The large capacitor C_f acts as a filter protecting the qubit from noise as explained in Sec. 2.3.2. **c** Theoretical fits to Eq. (2.41) for $V_{DC} = 0$. **d** From the fits of Eq. (2.41) ($V_p = 0$) to the hyperbolas showing a minimum, the mean parallel component of the TLS electric dipole moment relative to the electric field in C_{OV} is deduced to be 3 ± 1 D, which is comparable to other works [Bre+17; Sar+16].

outside of the measurement range in the PSS. Since PSS and VSS have been performed on different days it is possible that the contrast changes a bit so that TLS 3 being already barely detectable in VSS is not visible in the PSS at all. Second, as a result from coupling to a low-energy ($\sim k_B T$) and fluctuating TLS, the asymmetry energy of a probed TLS can switch irreversibly or back and forth on a time scale of hours to days. Thus, TLS may sometimes emerge or disappear in different data sets. The third possible explanation is that the TLS 1 to 4 very strongly respond to strain resulting in straight lines in PSS steeper than the V_p resolution. Further, there is an artifact of the swap-spectroscopy protocol represented by the dark and strongly tilted line at $V_p = -5$ V. The qubit excitation via the applied π-pulse fails when in resonance to a strongly coupled TLS. When such a TLS is tuned away by strain, it still can spoil the resolution due to a non-negligible probability of a Landau-Zener transition when the qubit is tuned through the TLS resonance by the swap pulse.

In the VSS, we can observe pronounced hyperbola that we interpret as sample-TLS hosted in the first OV. From fits to Eq. (2.41), the electric dipole moment components $d_{\parallel}$ parallel to the qubit field are deduced from Eq. (2.42) and shown in Fig. 4.3 (a) (blue). Since the qubit field direction within the OV coincides with that induced by the applied V_{DC}, the coupling strength g can be deduced from $d_{\parallel}$ using Eq. (2.34). At the symmetry point ($\varepsilon = 0$), TLS couple strongest to the qubit, the corresponding g values are presented in Fig. 4.3 (b) (blue). We see that the $d_{\parallel}$ values are similar to those observed in a 50 nm thick AlO_x sample formed via anodic oxidation [Bre+17], as well as in a 125 nm thick Si_3N_4 layer [Sar+16] formed in a plasma-enhanced chemical vapor deposition.

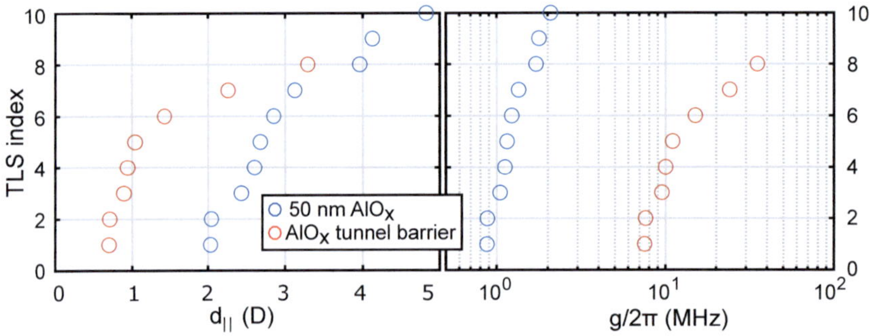

Figure 4.3: **a** TLS dipole moment components $d_{\parallel}$ parallel to the qubit field. Blue dots represent $d_{\parallel}$ values of TLS hosted in the 50 nm thick AlO_x sample dielectric (sample-TLS) that result from fits of Eq. (2.41) to the hyperbolas in Fig. 4.2 (d). The TLS presumably hosted in the tunnel barrier of the Josephson junction (junction-TLS) are represented by red dots. Their minimum $d_{\parallel}$ values are estimated from the level splittings in the qubit spectrum, assuming the likely case of near-symmetric TLS. **b** Measured coupling strengths g of junction-TLS (red), and deduced maximum g from $d_{\parallel}$ of sample-TLS (blue). The sample-TLS dipole moments are in good accordance to results obtained in other works [Bre+17; Sar+16]. Furthermore, they seem to be roughly three times larger than those of junction-TLS, which sounds reasonable since the Josephson junction's tunnel barrier is an interfacial thin film rather than a bulk dielectric. It is also possible that junction-TLS show effective and small dipole moments since they are screened by Cooper-pairs whose evanescent wave-function is non-zero in the tunnel barrier.

In five of six qubits, some avoided level crossings were observed using continuous wave spectroscopy that we refer to strongly coupled TLS. The corresponding coupling strengths g (half the level splitting) are shown in Fig. 4.3 (b) (red). If those TLS were located in the OV or at the circuit surface, their dipole moments would exceed 9 D and 30 D, respectively. This is however at least twice as large as the maximum $d_{\|}$ value deduced for sample-TLS, and close to a maximum imaginable TLS dipole moment in solids (10 D). Thus, we assume them to be located in the $\sim$2 nm thick tunnel barrier of the JJ. The TLS asymmetry cannot be deduced from avoided level crossings, thus minimum $d_{\|}$ values deduced from Eq. (2.34) are shown in Fig. 4.3 (a) (red) assuming the likely case of near-symmetric TLS. For comparison, we regard the splitting distribution observed with a phase qubit, presented in figure 2 (b) of the report by Martinis et al. [Mar+05], where the minimum and the maximum splittings correspond to TLS having $d_{\|}$ $\sim$0.6 D and $\sim$6 D, respectively. The here-detected dipole moments of junction-TLS are comparably distributed. However, their maximum $d_{\|}$ $\sim$3.2 D is smaller which is possibly due to the double-angle fabricated qubit JJs used here that may have a more clean tunnel barrier compared to the samples used in the work [Mar+05] where the JJ bottom electrode is argon-milled prior to the tunnel barrier formation.

Due to poor statistics, we only carefully note that the mean TLS dipole moments measured in 50 nm AlO_x are about three times larger than those of junction-TLS. This could indicate different microscopic origins of the compared TLS. While the sample-TLS are probably formed from tunneling ions, junction-TLS could additionally stem from electronic states at the metal-dielectric interface [FI06; Cho+09; Aga+13]. On the other hand, junction-TLS could be screened by Cooper-pairs whose evanescent wave function penetrates the tunnel barrier. This possibly naive explanation could be verified in TLS-sensors with very thin AlO_x sample films.

To summarize, we demonstrated the proof-of principle that the fabricated and measured first version of qubit-based TLS-sensors is suitable for exploring individual TLS in sample dielectrics. Especially, it enables characterization of TLS by strain- and E-field tuning, and even more importantly, this sensor architecture allows for distinction of TLS residing in the sample dielectric from the qubit's intrinsic TLS. Although this version shows a poor TLS resolution due to short energy relaxation time, it allows one to determine the sample dielectric loss tangent in good accordance to results from comparable systems based on microwave resonators. Further, we found hints that dipole moments of TLS hosted in the 50 nm AlO_x sample dielectric are about three times larger than those of TLS hosted in the qubit Josephson junction. This indicates a fundamental difference between TLS hosted at thin interfaces and those in rather thick dielectrics to be further investigated with the next generation of TLS-sensors. Due to time limitations, only their fabrication is reported in this dissertation (see Sec. 3.2.3 and (3.2.4)).

4.2 TLS detection in standard Xmons

Exploration of TLS in state-of-the-art qubits is a means to highlight weak-points of individual circuits regarding the TLS-generated energy loss and decoherence. Results of two developed methods to classify TLS and find their possible locations in Xmon samples [Bar+13], employing the interplay of dc-electric fields and static strain, are reported here. In the respective sections 4.2.1 and 4.2.2, the simulation method and results on electric fields at qubit interfaces generated by global gates are presented. Section 4.2.3 contains parameters of qubit samples probed in this work. In Sec. 4.2.4 experimental results on distinguishing junction-TLS are reported while in Sec. 4.2.5 the method to find possible locations of interface-TLS is detailed.

4.2.1 Simulation method of E-fields at qubit film edges

The simulation method reported here enables one to obtain the distribution of dc-electric fields induced by global electrodes which were used in the two experiments introduced in sections (2.3.4) and (2.3.5). As shown in the next section, the maximum electric field induced by the qubit decays approximately as $1/\sqrt{x}$, where x is the distance from the substrate-metal-vacuum edge, defined in the inset of Fig. 2.10 (b). A TLS with a maximum possible electric dipole moment of 10 D (2 eÅ) that resides at $x > 200\,\mathrm{nm}$ would couple to the qubit by $g/2\pi < 0.05\,\mathrm{MHz}$, which is the typical TLS detection sensitivity of the UCSB samples used in this work. It is thus sufficient to evaluate dc-fields, which are generated via global electrodes, within a narrow region $x < 300\,\mathrm{nm}$ at the qubit film edge, since TLS residing outside of this region practically do not couple to the qubit.

Three-dimensional model

The dc-electric fields were simulated using the electrostatic solution type of the finite element solver ANSYS Maxwell 2015 (release 16.2.0). Figure 4.4 (a) contains a cut view of the three-dimensional (3D) simulation model of the sample holder interior introduced in Sec. 3.3.3. The top inset shows the empty sample box, where the excavation in the center of the box floor models the empty space for the zircon sphere placed under the chip. The bottom inset contains a zoomed in side-view while the vertical transparent view in (b) indicates the global electrode's dimensions as well as the qubit positions on the UCSB sample chip [Bar+13] investigated with this setup.

Two-dimensional model

Since the qubit film thickness of 100 nm is by a factor $\sim 10^4$ smaller than the maximum global electrode dimension, the field simulation along the film edge profile would require an exorbitant amount of memory in the full 3D model. For this reason, the geometry is mapped onto a two-dimensional (2D) problem, where the top and bottom electrodes were replaced by effective ones at distances $d = 100\,\mu\mathrm{m}$ and $-50\,\mu\mathrm{m}$ relative to the qubit film, respectively (see Fig. 4.5 a). Hereby, the electric potential $\widetilde{V}_{t/b}$ of each effective electrode is deduced from two

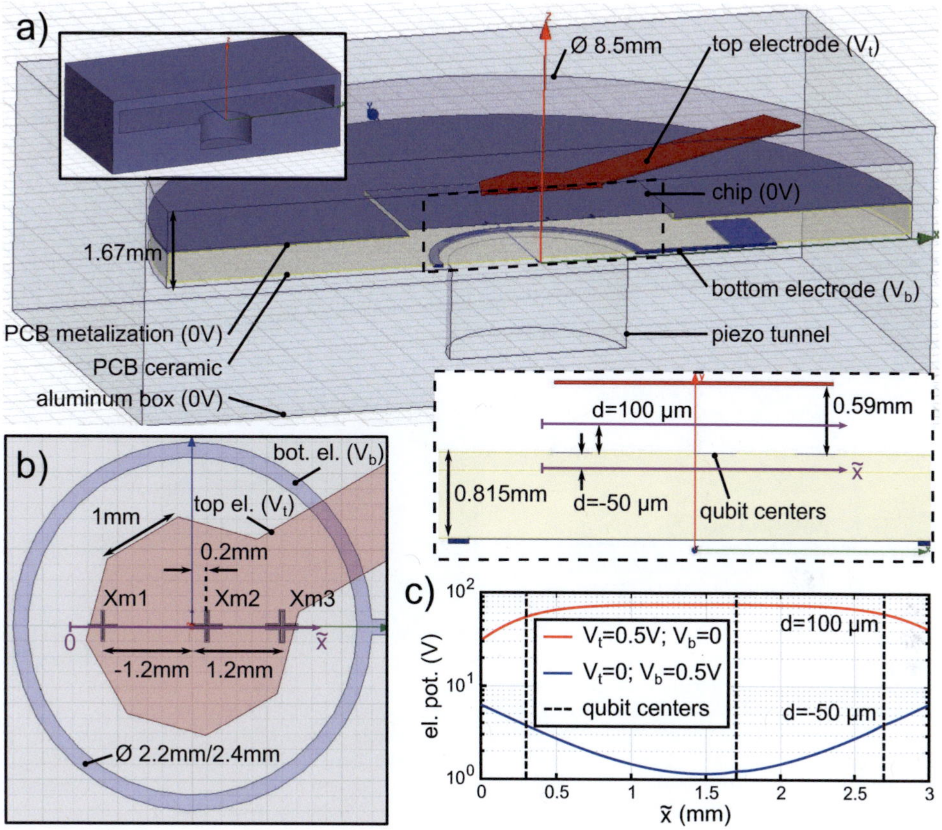

Figure 4.4: **a** Cut view of the three-dimensional (3D) model of the sample box interior introduced in Fig. 3.9 (c), used for coarse simulations of the dc-electric field induced by the top/bottom electrodes. The top and bottom insets show a cut view of the empty sample box, and the magnified side-view of the electrodes, respectively. **b** Transparent top view of the global electrodes and the three Xmon qubits (crosses) on the UCSB sample chip [Bar+13] investigated here. **c** Due to large computational effort, the 3D model was reduced to a two-dimensional (2D) one shown in Fig. 4.5, where the top/bottom electrodes at electrical potentials $V_{t/b}$ were replaced by effective top/bottom electrodes at potentials $\widetilde{V}_{t/b}$. To find the values $\widetilde{V}_{t/b}$, two 3D simulation runs were performed to find the electric potential along two straight lines shown in the bottom inset of (a) (violet), which are at the same distance to the chip metalization as the effective electrodes in the 2D model. Hereby, in each run a combination of voltages $V_{t/b}$ indicated in the legend was used. For each qubit, the pairs of values $(\widetilde{V}_t, \widetilde{V}_b)$ are given by the cross-sections with the dashed lines marking the qubit centers. The simulated field from the 2D model is employed in the method for detection of TLS locations in the qubit circuit as introduced in sections (2.3.4) and (2.3.5), and reported in Sec. 4.2.4 and (4.2.5).

separate simulation runs using the 3D model with a coarse resolution. In each run, either V_t or V_b were set to 0.5 V while the opposite electrode, the sample box, the printed circuit board (PCB) metalization and the chip ground plane were at 0 V. Figure 4.4 (c) contains a plot of the electric potential along two straight lines placed at same distances d to the chip metalization as the effective electrodes, indicated in the bottom inset of (a) (violet). The pair of values $(\widetilde{V}_t, \widetilde{V}_b)$ for each qubit is then deduced from cross-sections with black dashed lines marking the qubit centers. In other words, the dc electric potential of an effective electrode equals the one induced at its own location by the respective global electrode. Hereby, the ground plane almost fully screens the global electrodes from each other, so that corrections to the potential of the effective bottom electrode induced by the top electrode and vice versa are negligible. Further, at the eccentric locations of Xmon 1 and 3, this method becomes inaccurate since $\widetilde{V}$ varies on a comparable space scale as the dimension of the qubit island. The resulting field inhomogeneity is quantitatively captured with simulation data presented in Fig. 4.6 (b) to be discussed later.

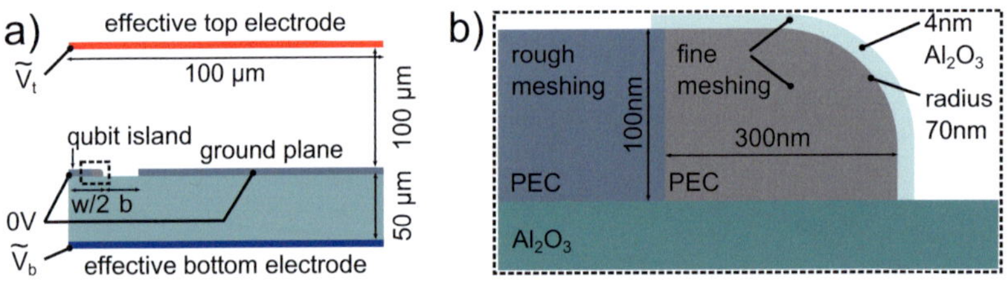

Figure 4.5: a Simplified two-dimensional (2D) model used to simulate the electric fields at the film edge profile of Xmon qubits [Bar+13] investigated in this work. Here, the film edge bulk is reduced to a 2D sheet, and the global electrodes are replaced by effective ones, whose electric potentials $\widetilde{V}_{t/b}$ are deduced from simulations of the coarse three-dimensional model (Fig. 4.4), as explained in the text. The qubit island potential is set to ground due to the transmon regime of the qubit. **b** Magnified view of the film edge profile. Only a small margin of the edge cross-section is fine resolved (maximum mesh width 0.1 nm) since the qubit fields are concentrated at the film edge, and TLS residing further away are not detectable by the qubit. The rest model is automatically meshed with a maximum mesh width of 250 nm. The aluminum film is modeled as a perfect conductor (PEC), and the amorphous native oxide on top of the aluminum by a 4 nm thick sapphire.

The simplified 2D model has a plate capacitor geometry, as shown in Fig. 4.5 (a). For symmetry reasons, only half of the qubit island is considered in the model while stray fields outside of the effective electrodes are suppressed by zero padding of the simulation region. The superconducting (aluminum) qubit film is modeled as a perfect conductor (PEC). Due to the transmon regime, the dc-electric potential of the qubit island is set to zero similar to the ground plane. To minimize calculation effort, only a 300 nm wide piece of the edge shown in Fig. 4.5 (b) is fine-resolved with a maximum mesh width of 0.1 nm. The rest island and ground plane have a maximum mesh width of 25 nm while the substrate, vacuum and both global electrodes are automatically meshed, which results in a mesh width scaling from 25 nm up to 250 nm at the global electrodes. As shown in Fig. 4.5 (b), the top corner of the film edge is modeled round (70 nm bend radius) due to dry etching used in the sample fabrication. Further, the metal is covered with a 4 nm thick sapphire layer that models the amorphous native oxide of the aluminum film. Since the finite element solver did not allow for such a small bend radius, the 2D model was scaled by a factor of 1×10^3 (nm, µV → µm, mV). All numbers given in this manuscript have the unchanged scale.

Discussion

Usage of the 2D model is justified by the assumption of a homogeneous dc-field, which is the case for the Xmon 2 qubit, as shown in Fig. 4.6 (b), where the deviation of the simulated ratio $E_t(y)/E_b(y)$ relative to the mean value is plotted along a path surrounding the qubit island illustrated in (a). Hereby, $E_{t/b} \equiv |\boldsymbol{E}_{t/b}|/V_{t/b}$ is the simulated electric field per applied voltage to the top/bottom electrodes. The data was obtained from a 3D simulation, where the qubit island and the ground plane were modeled by two-dimensional sheets with a relatively rough meshing (4 µm mesh width) to save calculation time. This explains the high roughness of the data plot which however is sufficient to verify the field homogeneity along the Xmon 2 film edge, and show the opposite for Xmon 1 and 3. Such behavior is explained by the eccentric location of Xmon 1 and 3 shown in Fig. 4.4 (b). The E_t/E_b variation for Xmon 1 is expected to be similar to that of Xmon 3, which was not explicitly verified due a finer required meshing and correspondingly high simulation time, related to its narrow geometry (see table of Fig. 4.6). For complicity, plots of $E_{t/b}(y)$ variations are shown in the appendix, Fig. F.6. The field inhomogeneity will be further discussed in Sec. 4.2.5.

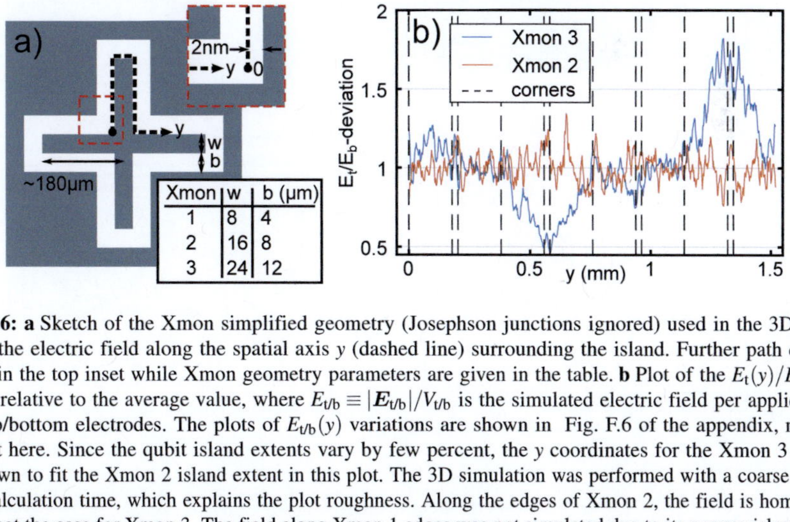

Xmon	w	b (µm)
1	8	4
2	16	8
3	24	12

Figure 4.6: **a** Sketch of the Xmon simplified geometry (Josephson junctions ignored) used in the 3D model to simulate the electric field along the spatial axis y (dashed line) surrounding the island. Further path details are depicted in the top inset while Xmon geometry parameters are given in the table. **b** Plot of the $E_t(y)/E_b(y)$-ratio variation relative to the average value, where $E_{t/b} \equiv |\boldsymbol{E}_{t/b}|/V_{t/b}$ is the simulated electric field per applied voltage to the top/bottom electrodes. The plots of $E_{t/b}(y)$ variations are shown in Fig. F.6 of the appendix, not further important here. Since the qubit island extents vary by few percent, the y coordinates for the Xmon 3 data were scaled down to fit the Xmon 2 island extent in this plot. The 3D simulation was performed with a coarse resolution to save calculation time, which explains the plot roughness. Along the edges of Xmon 2, the field is homogeneous, which is not the case for Xmon 3. The field along Xmon 1 edges was not simulated due to its narrow island and high required simulation time, however it is expected to have a similar form as that of Xmon 3. The field inhomogeneity makes the method of finding TLS locations in Xmon 1 and 3 unfeasible as further explained in Sec. 4.2.5.

4.2.2 Simulation results

For each qubit, three data sets were obtained from simulations using the two-dimensional (2D) model: two sets for the field $\boldsymbol{E}_{\mathrm{t/b}}(\boldsymbol{x})$ generated either by the top or the bottom electrodes, and one for the field $\boldsymbol{E}_{\mathrm{q}}(\boldsymbol{x})$ induced by the qubit oscillations. In the experiment on finding the TLS location in the qubit circuit introduced in Sec. 2.3.5, the simulated field is considered at four interfaces shown in Fig. 2.9. While the electric field direction and magnitude at all interfaces of Xmon 2 is presented in Fig. 2.10, some selected data at the interface between the native oxide and vacuum (OxV) are shown in Fig. 4.7 to visualize the influence of the qubit geometry and its position. A color plot of $|\boldsymbol{E}_{\mathrm{b}}(\boldsymbol{x})|$ for Xmon 2 is shown in the top inset of Fig. 4.7 (a), where the spatial axis leading along the OxV interface is depicted. The electric field generated by the bottom electrode at the OxV interface is the subset $\boldsymbol{E}_{\mathrm{b}}(x^{(\mathrm{OxV})})$ from the full data set $\boldsymbol{E}_{\mathrm{b}}(\boldsymbol{x})$. The figure 4.7 (a) contains the field magnitudes $|\boldsymbol{E}_{\mathrm{b/t}}(x^{(\mathrm{OxV})})|$ at the OxV interface for each qubit (see legend in (b)). The qubit arrangement relative to the global electrodes is sketched in the bottom inset of (a).

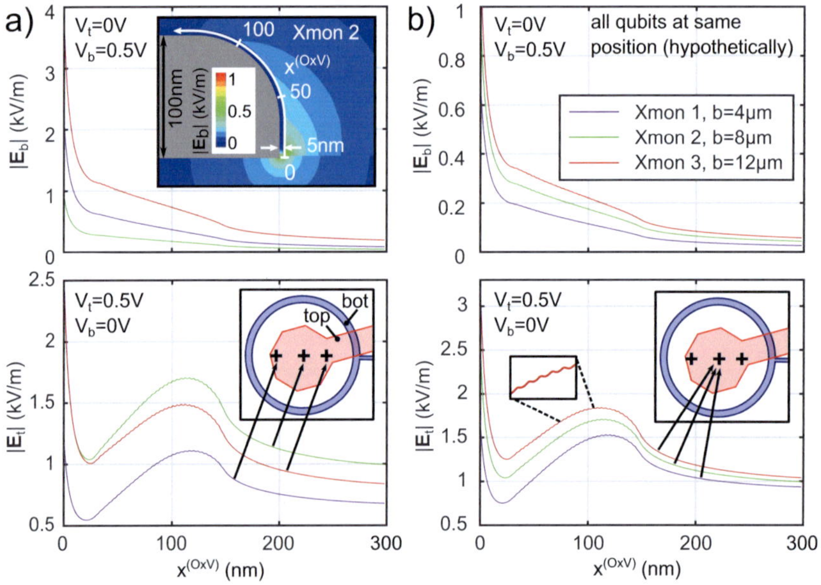

Figure 4.7: The electric field $\boldsymbol{E}_{\mathrm{b/t}}$ generated by the global bottom/top electrodes has been simulated using the two-dimensional model shown in Fig. 4.5. Here, $\boldsymbol{E}_{\mathrm{b/t}}$ is plotted along the spatial axis leading along the OxV interface (native oxide - vacuum), defined in the top inset of (a). Electric fields at other interfaces are shown in Fig. 2.10 (b) and (c). **a** Electric field magnitude $|\boldsymbol{E}_{\mathrm{b/t}}(x^{(\mathrm{OxV})})|$ induced by the bottom/top electrodes at voltages ($V_{\mathrm{b/t}} = 0.5\,\mathrm{V}, V_{\mathrm{t/b}} = 0\,\mathrm{V}$). The field strength at every qubit is influenced by its location relative to the global electrodes, illustrated in the bottom inset. **b** $|\boldsymbol{E}_{\mathrm{b/t}}(x^{(\mathrm{OxV})})|$ for the hypothetical case when all qubits have the location of Xmon 2 (see bottom inset). The data shown in (b) is not used in the analysis on TLS location in the qubit circuit, however it demonstrates that the applied dc-electric field increases with the increasing gap b between the qubit island and the ground plane listed in the legend. This behavior is worth mentioning, since the qubit-induced field evolves contrarily. Other geometrical details are listed in the table of Fig. 4.6. The curly line shape (see zoomed view) is due to meshing precision of the geometry, as further explained in the text.

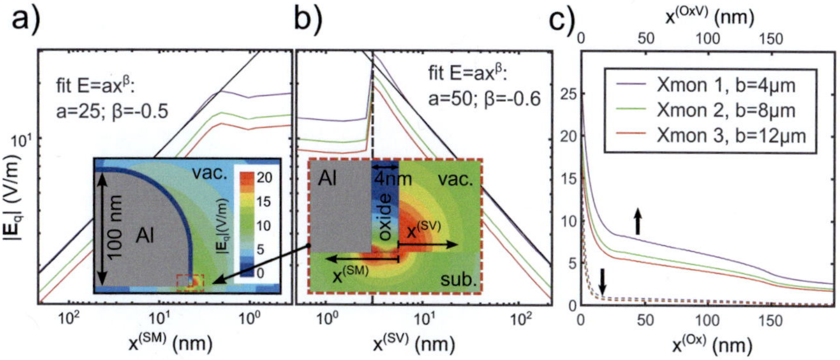

Figure 4.8: The electric field $\boldsymbol{E}_\mathrm{q}$ that is induced by the qubit oscillations has been simulated using the simplified two-dimensional model shown in Fig. 4.5: the dc electric potential of the effective electrodes is set to zero while that of the qubit island is $U_\mathrm{qub}^\mathrm{rms} \sim 5\,\mu\mathrm{V}$ which is the root mean square of the oscillating voltage between the transmon electrodes and is equal for all three Xmon qubits on the UCSB sample chip investigated here. $\boldsymbol{E}_\mathrm{q}$ is used to calculate the coupling strength g between the qubit and the investigated TLS, used in the analysis of TLS locations in Sec. 4.2.5. $|\boldsymbol{E}_\mathrm{q}|$ is plotted here along spatial axes (see inset of Fig. 2.10 b) defining the investigated interfaces. **a, b** Double-logarithmic plot of $|\boldsymbol{E}_\mathrm{q}|$ for each qubit (see legend in c), at the substrate-metal (SM) and substrate-vacuum (SV) interfaces, respectively. The black dashed line in **b** denotes the origin of the $x^{(\mathrm{SV})}$-axis. The fits indicate the polynomial decay law of the field strength quoted in the work by Barends et al. [Bar+13]. **c** $|\boldsymbol{E}_\mathrm{q}|$ in the native oxide layer of the aluminum film (bottom axis) and at the interface of the native oxide to vacuum (top axis). Comparison to Fig. 4.7 (a) emphasizes that the dc-electric field generated by the global electrodes increases, while the qubit-induced field decreases for a wider gap b.

The Xmon geometry parameters comprise the qubit island width w and the gap b between the island and the ground plane, which are listed in the table of Fig. 4.6. The cross' arm length is about 180 μm and not important for the 2D simulations. The data shown in Fig. 4.7 (b) are not used in this work, but serve to emphasize that the electrode-generated field at film edges increases with increasing b. These data sets were obtained when all qubit positions were set to that of Xmon 2 (see inset of c). All data $|\boldsymbol{E}_{\mathrm{b/t}}(x^{(\mathrm{OxV})})|$ and $|\boldsymbol{E}_{\mathrm{b/t}}(x^{(\mathrm{Ox})})|$ (not shown here) are curly in the region $0 < x^{(\mathrm{OxV})} < 150\,\mathrm{nm}$ and $0 < x^{(\mathrm{Ox})} < 150\,\mathrm{nm}$, respectively, as visible in the zoomed in view in (b). This is due to the simulation tool that approximates the curved top corner of the film edge by a triangular mesh, resulting in marginally increased field at each outstanding corner.

The qubit field at film edges was simulated with the 2D model, where the effective global electrodes and the ground plane were set to 0 V while the qubit island potential was set to $U_\mathrm{qub}^\mathrm{rms} = \sqrt{\hbar\omega_{01}/2C_\Sigma} \approx 5\,\mu\mathrm{V}$, the root mean square of the voltage induced by the qubit plasma oscillations in the qubit capacitor. ω_{01} is the qubit resonance frequency, and C_Σ the qubit total capacitance, both measured in this work and reported in Sec. (4.2.3). Figure 4.8 contains $|\boldsymbol{E}_\mathrm{q}|$ for each qubit (see legend) and each interface. The $\sim 1/\sqrt{x}$-dependence of the electric field along the substrate-metal (SM) and substrate-vacuum (SV) interfaces is visible due to the double logarithmic plots in (a) and (b) (see fits). Since the native oxide layer (Ox) and the OxV interface lead along the curved film top edge, there the field obeys this law piece-wise only. Comparison of Fig. 4.7 (a) with Fig. 4.8 points out that the electric field generated by the global electrodes increases, while the qubit-induced field decreases for a wider qubit gap b.

It is important to emphasize that for $x > 200\,\text{nm}$, the qubit field is below the needed value to detect a TLS with a maximum imaginable dipole moment of 10 D. For $x < 200\,\text{nm}$ however, the dc-electric fields induced by the global electrodes are strong enough to notably tune such a TLS, as quantitatively explained in Sec. 2.3.4. Thus, at all relevant interfaces (SV, SM, Ox, OxV), each detectable TLS should be dc-field tunable, which is not the case inside of the Josephson junction tunnel barrier where no dc-fields are induced. This principle to identify TLS hosted in the Josephson junction is used in Sec. 4.2.4.

4.2.3 Qubit parameters of the UCSB sample

The measured qubit parameters of the UCSB sample are listed in the table 4.2, where ω_{01} is the qubit resonance frequency, α the qubit anharmonicity defined in Eq. (2.21) and C_Σ the qubit total capacitance deduced from the transmon relation $-\hbar\alpha \simeq E_C = e^2/2C_\Sigma$. $T_{1,\text{Q}}$ is the typical qubit energy relaxation time whose maximum value has been observed to vary by up to 15% from one cool down to the next. The resonance frequency f_res and the loaded quality factor Q_L of the readout resonators have been obtained in its bright state [Ree+10].

qub. No.	$\omega_{01}/2\pi$ (GHz)	$\alpha/2\pi$ (MHz)	C_Σ (fF)	$T_{1,\text{Q}}$ (μs)	$g_\text{q}/2\pi$ (MHz)	$\omega_\text{res}/2\pi$ (GHz)	Q_L (10^3)
1	6.25	-220	88	2.1	24.4	6.38	53
2	6.30	-226	86	8.3	39.0	6.49	52
3	6.355	-220	88	7.3	32.4	6.59	80

Table 4.2: Xmon qubit parameters of the UCSB sample chip [Bar+13]. The one-photon resonance frequency ω_{01} and the qubit anharmonicity $\alpha = \omega_{12} - \omega_{01}$ are detected using continuous wave spectroscopy while α is twice the distance between the one-photon-driven 01 and the two-photon-driven 02 resonance peaks of the qubit. The total qubit capacitance C_Σ defined in Eq. (2.17) is calculated from the relation $\hbar\alpha \simeq -E_C = -e^2/2C_\Sigma$ valid in the transmon regime. These parameters are important to simulate the electric field at the qubit film edges induced by the qubit, which is required to deduce the coupling strength g of each detected TLS investigated in Sec. 4.2.5. The qubit energy relaxation time $T_{1,\text{Q}}$ results from a fit to the decaying population of the qubit excited state recorded with the time domain protocol shown in Fig. 2.3 (b). $T_{1,\text{Q}}$ is used to estimate the minimum detectable g with the corresponding qubit. The qubit-resonator coupling strength g_q is deduced using Eq. (2.19). The resonance frequency ω_res and the loaded quality factor Q_L are detected in the readout resonators many-photon state. The resonator internal quality factor Q_i in the few and many-photon state is deduced from quadrature fits to the test-resonators, and is $\sim 3\times10^5$ and $\sim 8\times10^5$, respectively.

4.2.4 Spectroscopy with applied elastic and electric fields: distinguishing junction-TLS

The method introduced in Sec. 2.3.4 to distinguish TLS hosted in the tunnel barrier of a Josephson junction (junction-TLS) from TLS residing at other qubit interfaces (interface-TLS) has been applied to Xmon [Bar+13] samples (shipped from UCSB, USA), and reported in the following.

Measurement data

The sample housing allowing for application of elastic strain and electric fields to the sample chip is introduced in Sec. (3.3.3). A simplified sketch of the chip holder is shown in the inset of Fig. 4.9 (a) where the piezo actuator and the global gates are depicted that are used to control the elastic or electric fields via applied dc-voltages V_p or V_{tb}, respectively. TLS resonance frequencies have been recorded using the swap-spectroscopy protocol (see Sec. 2.3.1), that was repeated for stepwise ramped V_p or V_{tb}. Measurements for swept V_p and constant V_{tb} are called piezo swap-spectroscopy (PSS), and dc-field swap-spectroscopy (DCSS) otherwise.

The exemplary data set shown in Fig. 4.9 (a) was recorded in two PSS measurements and one DCSS measurement using the qubit Xmon 2. The color of every pixel encodes the averaged population probability P_1 (see color bar) for the initially excited qubit after being tuned via a swap pulse to a given frequency (vertical axis) for a duration $\tau = 5\,\mu$s. Every dark line thus represents one TLS while the white curves in (b) are fits to Eq. (2.44). The red line is the qubit frequency for zero swap amplitude, which is not maximum due to trapped vortices from stray magnetic fields in the cryostat.

The fit factor γ_p in Eq. (2.44) is proportional to the averaged TLS deformation potential that can be interpreted as the elastic TLS dipole moment coupling to the strain field. γ_{tb} is proportional to the electric TLS dipole moment projection onto the electric field. Both fit factors can be understood as the slope of the hyperbola far away from its minimum. Due to the random distribution of TLS elastic dipole moment magnitudes and their orientations, the γ_p factors obtained from the PSS data are broadly distributed, which is obviously not the case for fit factors γ_{tb} extracted from the DCSS data set: A notably large subgroup of detected TLS do not respond to electric fields induced by the global electrodes, which we refer to TLS hosted in the tunnel barrier of the Josephson junction, where no dc-electric field is induced.

From the data presented in Fig. 4.9, one thus can identify junction-TLS due to their non-tunability by the dc-electric field. Hereby, only TLS whose resonance appears in both strain- and E-field sweeps can be evaluated. We thus interleaved PSS and DCSS sections, each spanning a narrow voltage interval of roughly 5 to 10 V while the total swept voltage range is of the order of 200 V. This increased the probability for each detected TLS to be visible in at least one DCSS section. Such a full data set is shown in Fig. D.3 of the appendix.

Data analysis: TLS density

A practical measure that quantifies the reliability of a qubit used for quantum computation is the density ρ of detectable TLS, which is the average TLS number per frequency unit. The underlying data set shown in Fig. D.3 consists of N PSS sections interlaced with M DCSS

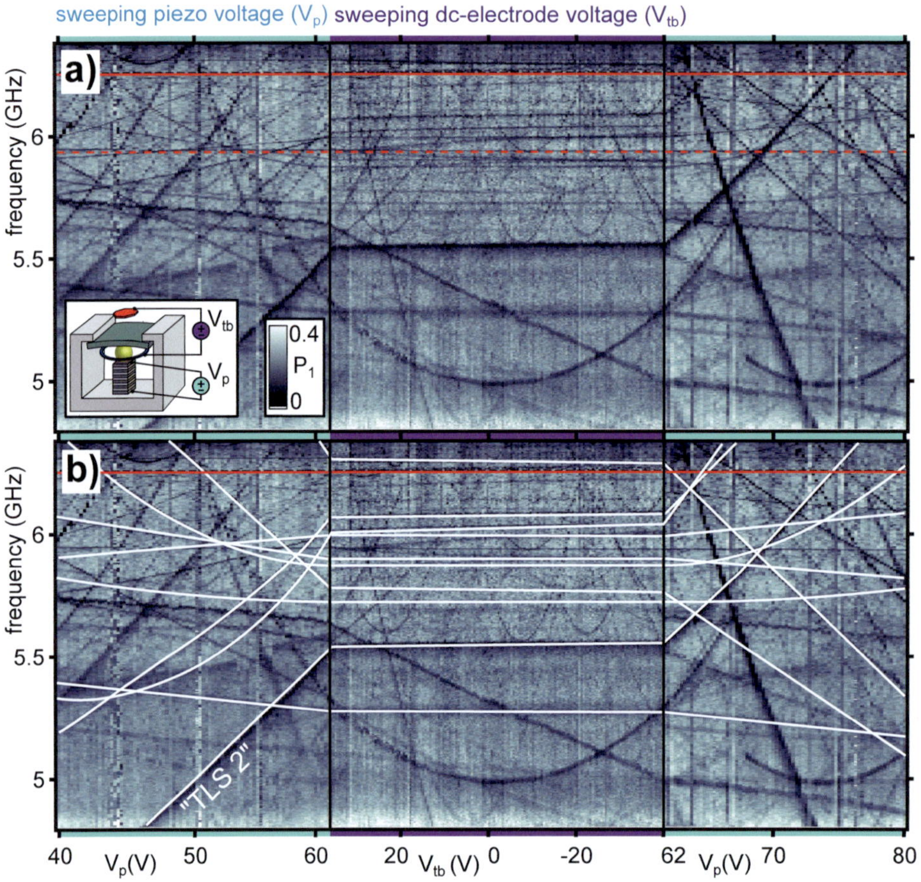

Figure 4.9: Data set recorded with Xmon 2, used to identify TLS hosted in Josephson junctions (JJs). **a** TLS distribution under elastic strain transferred via a piezo actuator or electric fields generated by global dc-electrodes (see inset). The TLS frequencies are recorded using the swap-spectroscopy protocol (see Sec. 2.3.1) repeated for stepwise ramped voltage V_p (cyan framed segments) or V_{tb} (violet). The color encodes the probability P_1 for the qubit to be in its excited state after an initial excitation and a swap to a given frequency (vertical axis) for a duration $\tau = 5\,\mu s$. The red line is the qubit frequency for zero amplitude of the swap pulse, which is not maximum due to background fields induced by trapped vortices. The red dashed line most probably indicates resonant cross-talk to another qubit on the chip. **b** Fits to equation Eq. (2.44) (white line) for a subgroup of TLS that do not respond to electric fields induced by global electrodes. Such TLS most probably reside in the tunnel barrier of the qubit Josephson junction (junction-TLS), where no voltage drop is induced due to the transmon regime. This method fails only in an unlikely case if a TLS electric dipole moment is perfectly parallel to the dc-field. The deduced TLS density of junction-TLS is about 15/GHz while it is 23/GHz for TLS hosted at other interfaces. Such high density of junction-TLS and especially their weak coupling to the qubit indicate that they predominantly reside in the stray junctions of the qubit, as further explained in the text.

sections, where N may deviate from M by one. The TLS density ρ_k in one of $N+M$ sections is calculated as

$$\rho_k = \eta_k \sum_i \sum_{V_k} \theta\left(\omega_{\text{TLS}}^{(i)}(V_k) - 2\pi f_{\min}\right) \theta\left(2\pi f_{\max} - \omega_{\text{TLS}}^{(i)}(V_k)\right), \tag{4.1}$$

where the first sum goes over all detected TLS, and the second sum goes through every voltage $V_k \in [V_{k,\min}, V_{k,\max}]$ applied in the section k. For example, in the left PSS of Fig. 4.9 (b), $[V_{k,\min}, V_{k,\max}]$ equals $[39, 62]$ V. The sum kernel is a product of two Heaviside functions, and equals one if the TLS frequency $\omega_{\text{TLS}}^{(i)}$ is inside the detection interval $2\pi[f_{\min}, f_{\max}]$, and zero otherwise. The normalization is

$$\eta_k = \frac{\Delta V}{V_{k,\max} - V_{k,\min}} \frac{1}{f_{\max} - f_{\min}}, \tag{4.2}$$

where ΔV is the voltage step that is equal for all PSS, and for all DCSS sections. Next, $V_{k,\max}^{(i)}$ and $V_{k,\min}^{(i)}$ are introduced that denote the minimum and maximum applied voltage in the section k, for which the TLS i is detectable. For example, in the left PSS of Fig. 4.9 (b), $V_{k,\max}^{(1)}$ and $V_{k,\min}^{(1)}$ of a "TLS 2" are 46 and 62 V, respectively. For a junction-TLS detected in a DCSS section, both values $V_{k,\max/\min}^{(i)}$ are trivially $V_{k,\max/\min}$. The expression in Eq. (4.1) thus simplifies to:

$$\rho_k = \frac{1}{V_{k,\max} - V_{k,\min}} \frac{1}{f_{\max} - f_{\min}} \sum_i V_{k,\max}^{(i)} - V_{k,\min}^{(i)}. \tag{4.3}$$

In words, ρ_k is obtained by counting the detectable TLS in the section k, each weighted with its occurrence in this section and normalized by the frequency window $f_{\max} - f_{\min}$. The total TLS density is the average over all ρ_k:

$$\rho = \frac{1}{N+M} \sum_k^{N+M} \rho_k. \tag{4.4}$$

Once junction-TLS are identified, the total TLS density ρ can be split into the junction-TLS density ρ_{J} and the density ρ_{I} of interface-TLS. In measurements with Xmon 2 and 3, about 100 TLS per qubit were detected. The deduced TLS densities are

$$\begin{aligned} \rho_{\text{J}} &= (15 \pm 1)/\text{GHz} \\ \rho_{\text{I}} &= (23 \pm 1)/\text{GHz} \end{aligned} \tag{4.5}$$

while the underlying raw data are shown in Fig. D.3 and Fig. D.4 in the appendix. Regarding the geometric parameters of the Xmon qubits listed in the table of Fig. 4.6, the mean edge length of the qubit island and the surrounding ground plane is ~3 mm. Thus, the density of interface-TLS along the Xmon film edge can be estimated to 7 /GHz/mm.

Discussion

In a recent measurement, J. Lisenfeld performed highly resolved swap-spectroscopy measurements, from which the qubit-TLS coupling strengths g were extracted by fitting Eq. (2.31) to the distinct local maxima of the qubit's energy relaxation rate. The corresponding g histograms of interface-TLS (field-tunable) and junction-TLS (non-tunable) are shown in Fig. 4.10 (a) and (b), respectively. The blue colored data have been obtained with Xmon 2 while the red bars represent data from Xmon 3. Surprisingly, the non-tunable TLS exhibit small g values that follow a similar distribution as the coupling strengths of interface-TLS. Assuming them to reside in the small Josephson junctions implies unusually small electric dipole moments $d_{\|} \in [3..25]\,\mathrm{mD} = [0.7..5] \times 10^{-3}\,\mathrm{e\AA}$.

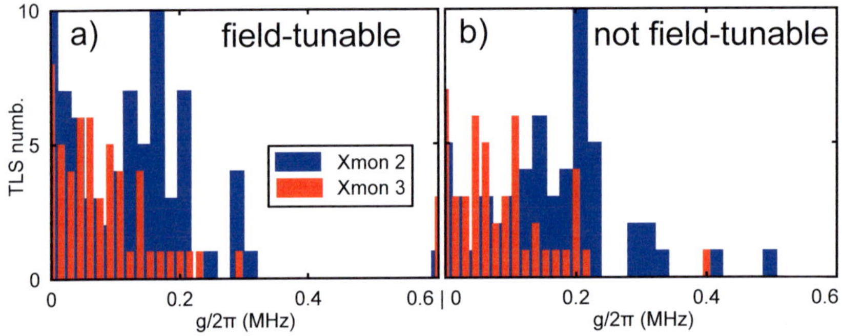

Figure 4.10: TLS-qubit coupling strengths g extracted from Lorentzian fits to peaks in the qubit's energy relaxation rate that was obtained with fine-resolved swap-spectroscopy. The minimum value is around $g/2\pi \approx 0.04\,\mathrm{MHz}$. **a** Histogram of g for TLS that respond to the dc-electric fields generated by the global electrodes. The legends indicate the used qubit sample. This data supports the outcome of the TLS location analysis reported in the next section, to be discussed there. **b** Histogram of g for TLS that are not tunable via the electric field. Both g distributions in (a) and (b) are similar, which supports the explanation that the non-tunable TLS reside in the stray Josephson junctions.

Most probably, the weakly coupling TLS that are not field-tunable reside in the stray junctions, as recently discussed with Anthony Megrant (Google inc.). As shown in the inset of Fig. 2.9 (a), the small and the stray junctions are connected in series, and act as a voltage-divider. Comparing their areas ($200\,\mathrm{nm} \times 300\,\mathrm{nm}$ and $3\,\mu\mathrm{m} \times 4\,\mu\mathrm{m}$) leads to electric field strengths induced by the qubit vacuum fluctuations of about 2.42 kV/m and 12 V/m inside the identically formed tunnel barriers of the small and the stray junctions, respectively. Together with recorded coupling strengths in Fig. 4.10 (b), the deduced dipole moments of the stray-junction-TLS range from 0.7 to 7 D which are reasonable numbers.

The location of non-tunable TLS in the Josephson junctions is further supported by the simulation results: no other location in the qubit circuit exists where the qubit fields are non-zero while the field generated by the global electrodes vanishes. In a seldom case, the dipole moment of a TLS can be oriented perpendicularly to the applied fields, which however would also decouple this TLS from the qubit.

Alternatively, the non-tunable TLS could stem from spins that reside close to or inside the small Josephson junction, and couple to magnetic fields induced by the qubit oscillating currents while being decoupled from electric fields. In a simplified picture, the coupling strength of the qubit to an electron spin $1/2$ residing at the film surface is maximally 50 Hz as deduced in the appendix E. This however is far below the minimum measured $g/2\pi \approx 40\,\text{kHz}$. Further, no offset magnetic fields are present at superconducting qubits which would raise the Zeeman splitting of a spin $1/2$ in resonance with the qubit. Equally, the hyperfine splitting of an S electron in hydrogen is too small ($\sim$1.4 GHz) and further decreases when regarding heavier atoms.

Summarizing, a method has been implemented to tune TLS with elastic strain and dc-electric fields, which allows one to identify TLS residing inside the tunnel barrier of the qubit's Josephson junctions. This technique does not require any special qubit geometries. It works with ready-made Xmon-type and floating transmon samples but also with qubits in the phase regime ($E_J \ggg E_C$). Independently of the E_J/E_C ratio, it is applicable to qubits whose topology is that of a ring interrupted by a Josephson junction, where both JJ electrodes are at same dc-electric potential since they are shorted by the ring. This experiment is currently in a publication process [Lis+19].

4.2.5 Deducing TLS positions with two dc-electrodes

As reported in the previous section, TLS hosted in the Josephson junction of a transmon qubit (junction-TLS) are characterized by their zero-response to applied dc-electric fields. Other detectable TLS presumably reside at film interfaces of the qubit circuit, and are called interface-TLS. The method to find locations of individual interface-TLS at the profile of qubit film edges is explained in Sec. 2.3.5. It has been applied to Xmon [Bar+13] samples that were shipped from UCSB, USA, as reported in the following.

Measurement data

While the piezo actuator was not used here, the global electrodes introduced in Sec. (3.3.3) were employed to generate dc-electric fields at qubit interfaces. As illustrated in the inset of Fig. 4.11 (b), two voltage sources were used to control the voltage drops $V_{t/b}$ from the top/bottom electrodes to ground, respectively. The TLS resonances were recorded by repetitive application of the swap-spectroscopy protocol (see Sec. 2.3.1) while iteratively ramping either V_t or V_b. The data set obtained for swept V_t and constant V_b is called top electrode swap-spectroscopy (TSS), and bottom electrode swap-spectroscopy (BSS) otherwise.

Figures 4.11 (a) and (b) contain exemplary segments of data sets obtained with Xmons 2 and 3, respectively. Since the Xmon 1 data look very similar to those obtained with Xmon 3, they are not shown here, but raw data are presented in the appendix (Fig. F.2 to F.4) for each qubit. The TSS data subsets are red margined, and the BSS subsets have a blue frame. The color of the surface plots encodes the averaged qubit excited state population P_1

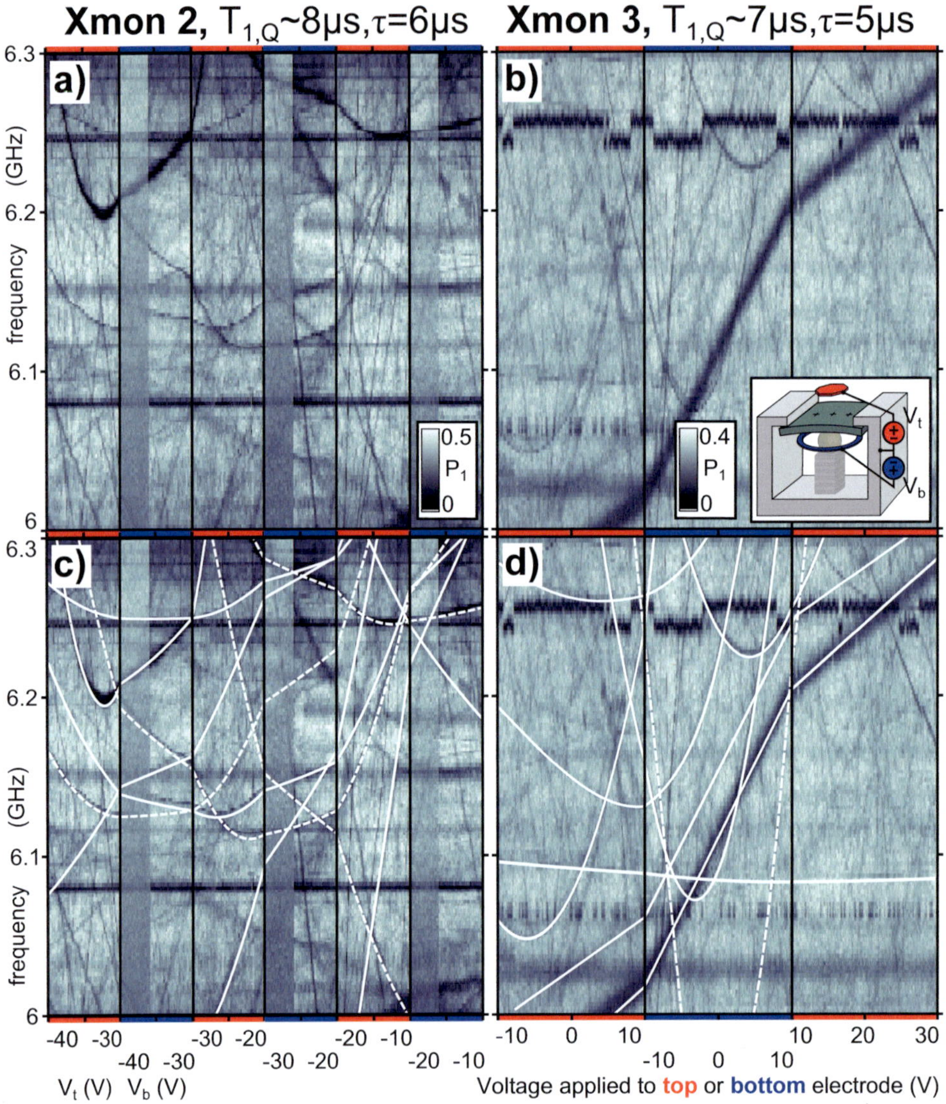

Figure 4.11: Segments of data sets recorded with Xmon 2 and 3 used for the analysis of TLS positions at qubit interfaces employing two external electrodes. TLS frequencies were recorded by swap-spectroscopy (swap duration τ) resulting in the remaining qubit excited state population P_1 (color-coded) vs. probing frequency. **a** Interleaved data sets recorded with Xmon 2 for swept V_t and constant V_b (red margin), and vice versa (blue margin). $V_{t/b}$ are voltages applied to the global electrodes inllustrated in the inset of (b). **b** Similarly obtained data with Xmon 3. Data from Xmon 1 are not shown here since they look comparable. The full data sets for each qubit are shown in the appendix (Fig. F.2 to F.4). **c, d** Hyperbolic fits of Eq. (2.47) to some exemplary TLS traces, indicated by white continuous and dashed lines for better readability. The extracted ratios γ_t/γ_b of the asymptotic tilts are presented in Fig. 4.12 (b) and further used in the TLS position analysis.

after excitation and subsequent tuning of the qubit to a varying probe frequency (vertical axis) for a duration τ. Each dark pixel means that the qubit has been near resonance to a TLS at the given frequency and given applied voltages (V_t, V_b). The horizontal dark lines are traces of junction-TLS that are ignored in this experiment. Since the generated electric fields from both electrodes are additive, V_t and V_b are swept consequently up- or downwards to improve statistics by increasing the number of TLS being tuned through the frequency tunability range of the qubit. In every BSS subset of (a), a blurred vertical stripe is visible that results from a failed qubit frequency calibration, which however does not influence the analysis.

Hyperbolic fits of Eq. (2.47) to exemplary TLS traces are shown in Fig. 4.11 (c) and (d), where some lines are dashed to preserve the plot clearness. The individual fit factors $\gamma_{t/b}$ are asymptotic hyperbola slopes, and are interpreted as the respective TLS tunabilities by the top/bottom electrodes. The set of ratios γ_t/γ_b extracted for each qubit is the starting point for the analysis of TLS positions introduced in Sec. 2.3.5.

Two TLS detected at ~6.35 GHz and ~6.07 GHz in (b) show random switching between two frequencies. This is most probably due to their respective longitudinal coupling to TLS with a transition frequency close to the thermal level $k_B T$, that randomly switch between their eigenstates. Such systems have been thoroughly studied in the work of Meißner et al. [Mei+18], and they most probably are the reason for resonance frequency and energy relaxation fluctuations observable in transmon qubits [Mül+14; Kli+18; Sch+19].

Data analysis: film interfaces

First, TLS positions at film interfaces are considered that comprise the substrate-metal (SM) interface, the native oxide layer on top of the aluminum film (Ox) and the interface at the native oxide and vacuum (OxV). Simulated ratios E_t/E_b are shown for each qubit in Fig. 4.12 (a), where the spatial axes at each interface are illustrated in the inset. Hereby, $E_{t/b} \equiv |\boldsymbol{E}_{t/b}|/V_{t/b}$ is the electric field strength generated by the top/bottom electrodes per voltage applied. The simulation method and results are reported in sections 4.2.1 and 4.2.2, respectively. Histograms of measured ratios γ_t/γ_b are shown in Fig. 4.12 (b), with the respective number of analyzed interface-TLS from one cryogenic run indicated in the legend. For each qubit and for each TLS, comparison of (a) and (b) delivers a set of possible positions $\{x\}$ at the film interfaces to be truncated as explained in the following.

The fit factors γ_t and γ_b obtained with Xmon 2 are plotted vs. the ratio γ_t/γ_b in Fig. 4.12 (c), where black reference lines interconnect data points of a same TLS. Similar plots from Xmon 1 and 3 are shown in the appendix (Fig. F.1). Combining $|\boldsymbol{E}_{t/b}(x)|$ as a function of deduced TLS positions $\{x\}$ with $\gamma_{t/b}$ in Eq. (2.48) leads to TLS electric dipole moment components $d_{\|t/b}$ parallel to the electric field generated by the top/bottom electrodes. Since $\boldsymbol{E}_t$ and $\boldsymbol{E}_b$ are parallel at film interfaces, $d_{\|t}$ equals $d_{\|b}$, as verified in the appendix, see Fig. F.5 (a). The deduced solutions $d_{\|t}$ are plotted in Fig. 4.12 (d). It is visible on the logarithmic scale that the dipole moment solutions based on locations $\{x^{(Ox)}\}$ are about ten times larger than those deduced from $\{x^{(OxV)}\}$ (more details in appendix, Fig. F.5 b). This is due to

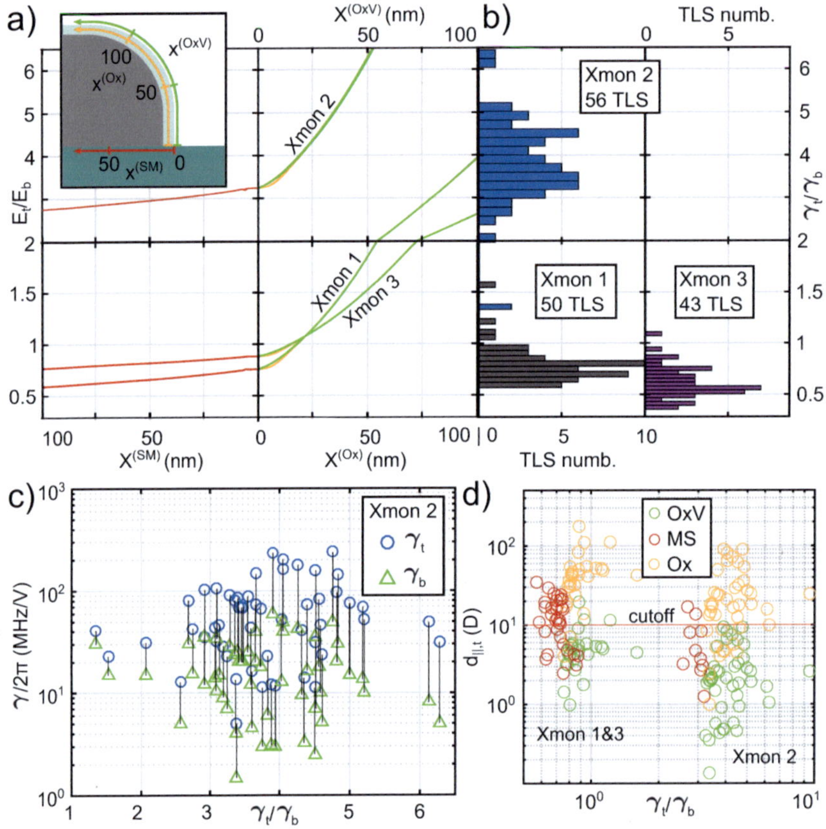

Figure 4.12: Evaluation of TLS positions at film interfaces (OxV, SM, Ox). **a** Ratio E_t/E_b for each qubit and each interface, along spatial axes indicated in the inset. Here, $E_{t/b} \equiv |\boldsymbol{E}_{t/b}|/V_{t/b}$ is the simulated electric field strength per voltage applied. At Ox and OxV interfaces, the E_t/E_b curves almost overlap and are slightly shifted for better visibility. Xmon 2 couples by roughly four times weaker to the annular bottom electrode than other qubits due to its central position. Therefore, the vertical axis is rescaled in (a) and (b), for clarity. **b** Histograms of individual TLS tunability ratios γ_t/γ_b obtained from fits to data shown in Fig. 4.11. One out-lier at $\gamma_t/\gamma_b \approx 9$ is not shown in (b) and (c), but in (d). For each TLS, comparison of (a) and (b) leads to a set of possible locations at every interface. **c** γ_t and γ_b recorded with Xmon 2, and plotted vs. the γ_t/γ_b ratio (see appendix, Fig. F.1 for Xmon 1 and 3 data) where black reference lines interconnect data points of a same TLS. **d** TLS electric dipole moment component $d_{\parallel t}$ parallel to $\boldsymbol{E}_t$, deduced from TLS position x, $E_t(x)$ and γ_t ($d_{\parallel t} = d_{\parallel b}$ at film interfaces). All position solutions requiring an unreasonably large dipole moment component $d_{\parallel t/b} > 10\,\mathrm{D}$ are ignored (red line). Solutions at the Ox interface are usually discarded due to a weaker electric field inside of AlO_x. Due to the eccentric position of Xmon 1 and 3 the field along their film edges is inhomogeneous (see Fig. 4.6 b). This method is thus applicable to Xmon 2 only, as further explained in the text.

the field strength inside the native oxide being by $\sim 1/\varepsilon_r \approx 0.1$ smaller than that at the oxide surface. The red line indicates a cutoff: since 10 D (2 eÅ) is a maximum imaginable value for an electric dipole moment in solids, all solutions implying $d_{\parallel \mathrm{t/b}} > 10\,\mathrm{D}$ are disregarded.

For ~9 % of interface-TLS detected with Xmon 2, no solutions are available after truncation, which sounds reasonable, since the SV interface has not yet been considered. In contrast, this number is 40 % and 86 % in the case of Xmon 1 and 3, respectively, which indicates that some assumptions made in this analysis do not apply to the latter qubits. Indeed, Xmon 1 and 3 experience an inhomogeneous dc-field due to their eccentric positions relative to the global electrodes: as simulated and shown in Fig. 4.6 (b), the $E_\mathrm{t}/E_\mathrm{b}$ ratio deviates by a factor of 0.5 to 1.8 along the island edge relative to the value when homogeneous field is assumed. Correspondingly, most $\gamma_\mathrm{t}/\gamma_\mathrm{b}$ ratios are compared with the wrong $E_\mathrm{t}/E_\mathrm{b}$ value resulting in wrong TLS positions. This method is thus inapplicable to Xmon 1 and 3, and the results presented in the following are valid for Xmon 2 only.

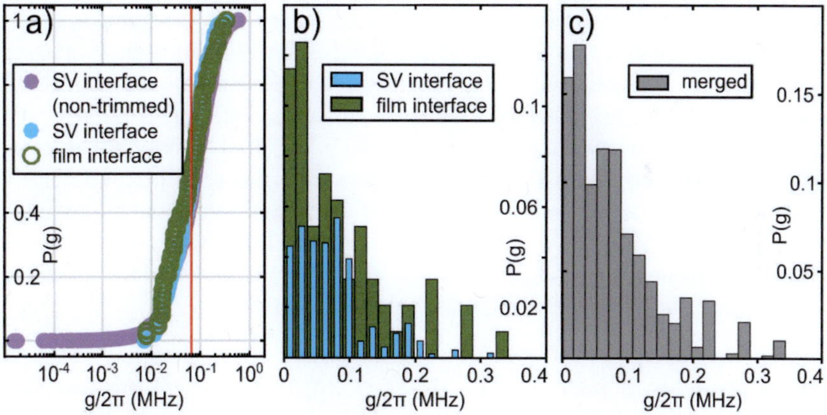

Figure 4.13: **a** Normalized cumulative distribution of the maximum TLS-qubit coupling strength g deduced from Eq. (4.6) when assuming a symmetric TLS ($\varepsilon = 0$). Green circles indicate g values after truncation, assuming the TLS to reside at film interfaces (OxV, SM, Ox). Violet and cyan dots indicate g assuming TLS to be hosted at the SV-interface, before and after truncation, respectively. The minimum $g/2\pi \sim 0.01\,\mathrm{MHz}$ is smaller than the predicted minimum detectable TLS-qubit coupling strength $g_\mathrm{min}/2\pi \approx 0.065\,\mathrm{MHz}$ (red line) which possibly stems from too strict assumptions further discussed in the text. **b** Binned g from the truncated data presented in (a). Each histogram is weighted with $P^{(SV)}$ (cyan) and $P^{(OxV)} + P^{(SM)} + P^{(Ox)}$ (green), where $P(g)$ is the relative accumulation (vertical axis). $P^{(i)}$ is defined in Eq. (2.51), and is the relative likelihood for an analyzed TLS to reside at the interface i. **c** Merged histogram valid for all analyzed interfaces, which reflects a similar distribution of g as that obtained from direct measurements, shown in Fig. 4.10 (a). This supports the correctness of the TLS-location analysis.

The coupling strength g between individual TLS and the qubit

$$\hbar g = \frac{\Delta}{E}\frac{\hbar \gamma_{\mathrm{t/b}} V_{\mathrm{t/b}}}{2E_{\mathrm{t/b}}(x)} E_\mathrm{q}(x) \tag{4.6}$$

is calculated from Eq. (2.48) and the identity $\hbar g = (\Delta/E)d_{\|,t/b}|\boldsymbol{E}_q(x)|$, where $\boldsymbol{E}_q$ is the simulated qubit field ($d_{\|,t} = d_{\|,b}$ at film interfaces). In figure 4.13 (a), green circles denote the normalized cumulative distribution of g deduced from TLS locations at film interfaces. Hereby, the investigated TLS were assumed to be symmetric ($\varepsilon = 0$) so that $\Delta/E = 1$, and g is maximum. The smallest $g/2\pi \sim 0.01$ MHz falls below the minimum detectable coupling strength 0.065 MHz (red line) calculated for Xmon 2 from Eq. (2.32), to be discussed later.

Data analysis: substrate-vacuum interface

After deriving the solution set for TLS locations at film interfaces, the SV interface is considered. Following the instructions in Sec. 2.3.5, for each TLS a set of possible $(x^{(SV)}, \alpha_d)$-tuples is deduced, where α_d implies the TLS electric dipole moment orientation, illustrated in Fig. 4.14 (c). The corresponding $d_{\|,t/b}$ and g solutions are deduced as previously while accounting in addition for α_d. Similarly, the SV-solutions that imply $d_{\|t/b} > 10$ D are ignored. In Fig. 4.13 (a), the normalized cumulative distribution of maximum g derived from SV-solutions before and after truncation is depicted by violet and cyan dots, respectively. The coupling strength distributions deduced from film and SV interfaces are similar to Monte Carlo simulation results shown in the work of Barends et al. [Bar+13]. Figure 4.13 (b) contains g histograms of truncated data shown in (a), that are normalized and weighted with $P^{(\mathrm{SV})}$ (cyan) and $P^{(\mathrm{OxV})} + P^{(\mathrm{SM})} + P^{(\mathrm{Ox})}$ (green), where $P(g)$ is the relative accumulation (vertical axis). The mean probability $P^{(\mathrm{i})}$ for an interface-TLS to reside at the interface i is presented in the following. The merged normalized g histogram shown in Fig. 4.13 (c) is similar to the coupling strength distribution (see Fig. 4.10 a) directly measured in a recent experiment. Further, taking into account the simulated qubit-field, the electric dipole moment components of interface-TLS deduced from Fig. 4.10 (a) range from 0.4 to 10 D, which is in good agreement with the dipole moments (Fig. 4.12 d) deduced with the method reported in this section. Both similarities support the correctness of the TLS-position analysis.

The deduced positions of interface-TLS detected with Xmon 2 are presented for every interface in Fig. 4.14 (a), and the respective spatial axes are indicated in (b). Each histogram is normalized with $P^{(\mathrm{i})}$ listed in the legend. The SV-histogram appears smoothed in comparison to the other data: at film interfaces, the solutions x are degenerate in $\alpha_d \in [0..\pi]$, hence, each TLS contributes by at most one solution to each histogram. However, for each TLS hundreds of solution combinations (x, α_d) are found at the SV interface, and merged in one histogram (cyan). Hereby, α_d was traced out, assuming the TLS orientation to be isotropic, which sounds reasonable due to the surface roughness of the SV interface, induced by micro-fabrication.

Error estimation

The precision of this method is dominated by the determination error of the distances $h_t = 590\,\mu$m and $h_b = 815\,\mu$m between the qubit film and the top and bottom electrodes, respectively. The h_b deviations may result from the unevenly machined ceramic between the chip and the bottom electrode while h_t uncertainty may stem from the stack of the isolation and Al foils underneath the printed circuit board (see illustration in Fig. 3.9 c).

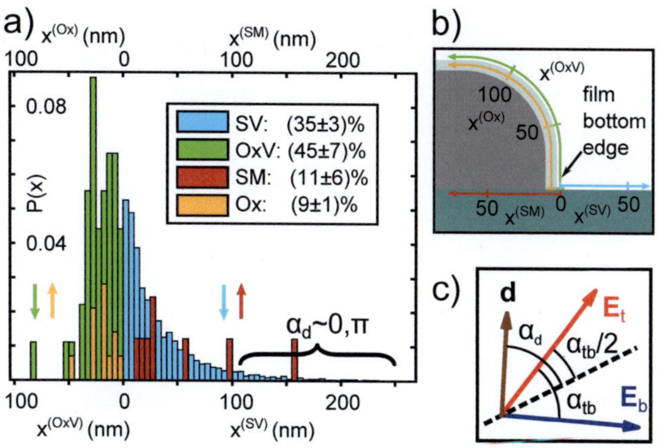

Figure 4.14: a Histogram of possible TLS positions x at interfaces investigated here and derived from the measured tunability of individual TLS by the top/bottom global electrodes. **b** Definition of the spatial axes for each interface. **c** The TLS electric dipole moment orientation relative to the dc-electric fields is given by the angle α_d. At film interfaces (OxV, SM, Ox), TLS positions are degenerate in α_d so that each TLS contributes by at most one entry in each histogram. At the SV-interface, however hundreds of solutions (x, α_d) are found per TLS, and merged in one histogram (α_d was traced out), which explains its smoothed appearance. The averaged probability $P^{(i)}$ for an interface-TLS to reside at the interface i is deduced from Eq. (2.51) and listed in the legend. The dominant uncertainty in determining TLS positions and $P^{(i)}$ is given by the estimated distances $h_{\mathrm{t/b}}$ of the qubit film to the top/bottom electrodes. The analysis has been repeated for varying $h_{\mathrm{t/b}}$ to deduce $P^{(i)}$ errors as explained in the text and with further details in the appendix (Fig. F.7). Most investigated interface-TLS seem to reside at the OxV and SV-interfaces, which may result from adsorbates, or chemical compounds from fabrication processes at the circuit surface, but also from residuals of a protecting resist used for sample shipping. The latter assumption is supported by the narrow distribution of TLS positions around the film bottom edge, which on the other hand could be due to the qubit field concentrated in this spot.

The analysis reported in Fig. 4.14 (a) has been repeated by varying $h_{\mathrm{t/b}}$ in discrete steps $\delta h = \pm 10\,\mu\mathrm{m}$, not exceeding the maximum imaginable deviation of $\pm 50\,\mu\mathrm{m}$. From each variation, the participation $P^{(\mathrm{i})}$ of the interface i is derived using Eq. (2.51). The numbers listed below are average values from these variations, and the according errors represent the standard deviations:

$$
\begin{aligned}
P^{(\mathrm{SV})} &= (35 \pm 3)\,\%,\\
P^{(\mathrm{OxV})} &= (45 \pm 7)\,\%,\\
P^{(\mathrm{SM})} &= (11 \pm 6)\,\%,\\
P^{(\mathrm{Ox})} &= (9 \pm 1)\,\%.
\end{aligned}
\tag{4.7}
$$

In Fig. F.7 of the appendix, the analysis outcome for a rough sweep step of $\delta h = \pm 25\,\mu\mathrm{m}$ is presented to give the reader an impression how the x-histograms and the interface participations change under distance variations. Further, the location analysis has been repeated for various cutoff values of the dipole moment, as reported in the appendix (Fig. F.8). The results support the notion that the here-applied cutoff of 10 D is reasonable since a larger dipole moment is hardly imaginable in solids, while smaller cutoffs lead to loss of information.

Discussion

Most investigated TLS seem to reside within a surprisingly small distance of 50 nm from the film bottom edge illustrated Fig. 4.14 (b). This observation results from the qubit energy relaxation time $T_{\mathbf{1,Q}}$ that is short compared to the nominal value $\sim$20 µs [Bar+13]. Short $T_{\mathbf{1,Q}}$ sets a bound for the minimum detectable coupling strength between qubit and TLS, so that merely TLS are resolved that reside in regions of highest qubit-induced fields which is at the film bottom edge. Besides BCS quasiparticles [Cat+11] and vortices [Son+09] that may limit $T_{\mathbf{1,Q}}$, another loss mechanism is suspicious that in addition may increase the TLS density at circuit surfaces, namely residuals from the protecting resist (SPR955) used for sample shipping. The microchip was covered by the resist for about two years and was cleaned in an acetone bath within a dozen of minutes. It would be interesting to know whether thoroughly cleaning the sample would alter the measurable TLS position distribution and reveal also defects in larger distances from the film edge. A possible, more sophisticated cleaning recipe is a bath in a 90 °C hot N-Ethyl-2-pyrrolidon (NEP) for 6 h. The assumption of a high degree of contamination from resist residuals is also supported by the observation that most investigated interface-TLS reside at the OxV and SV interfaces. On the other hand, the sample surfaces partially can be damaged from the etching plasma used for sample fabrication [Dun+17], they can be contaminated with OH groups from etching and post-fabrication processing [Mar+05; Sha+10; Gor+14], and they host frozen water and air when cooled down [Kum+16; Gra+17], which equally may lead to formation of TLS.

A small percentage of detected interface-TLS reside in the native oxide layer. This however does not exclude existence of TLS in the amorphous surface dielectric where the electric fields are so weak that a qubit energy relaxation time of $T_{1,\mathrm{Q}} > 100\,\mu\mathrm{s}$ is required to be sensitive to TLS hosted in the native oxide. Interestingly, 100 µs is roughly the limit of $T_{1,\mathrm{Q}}$ known in transmon qubits, see FIG 4 of the work by Wang et al. [Wan+15]. This may indicate that most coherent transmons are limited by dielectric losses induced in the native Al surface oxide.

As mentioned before, the minimum TLS-qubit coupling strength g deduced from locations of interface-TLS is below the Xmon 2 sensitivity g_{min} that was estimated in Eq. (2.32) assuming a TLS coherence rate of $\Gamma_{1,\mathrm{TLS}}/2+\Gamma_{\phi,\mathrm{TLS}} \approx 1/(100\,\mathrm{ns})$ [Bar+13]. $\Gamma_{1,\mathrm{TLS}}$ is the TLS energy relaxation rate and $\Gamma_{\phi,\mathrm{TLS}}$ its pure dephasing rate. Assuming $\Gamma_{1,\mathrm{TLS}}/2+\Gamma_{\phi,\mathrm{TLS}} \approx 1/(1.4\,\mu\mathrm{s})$ would lift this discrepancy, this however contradicts the recently measured TLS coherence times [Lis+19]. The discrepancy can be explained qualitatively: a TLS trace in data shown in Fig. 4.11 consists of dips in the recorded signal interpreted as P_1. The g_{min} estimation method premises that a single P_1 dip is identifiable as such. Since the TLS traces are fitted by hand, a hyperbola has a chance to be recognized by a human, although its single dips are barely identifiable. In other words, the field-response of weakly coupling TLS is more easily identified compared to single dips in P_1.

If Xmons are fabricated without so-called bandaging [Dun+17] on silicon wafers, dielectric losses are dominated by the films forming the Josephson junction (JJ), whose fabrication requires aggressive ion milling. The herewith induced structural damage of the substrate and contaminations lead to TLS formation in vicinity of the JJs, where the qubit field is increased [Dun+17]. This effect probably plays a small role for non-bandaged Xmon samples investigated here that were fabricated on a less fragile sapphire substrate. A possible experiment to verify this would be to employ an on-chip electrode placed close to the JJs, which would couple to TLS on junction leads in contrast to a global electrode that would equally affect all interface-TLS. This would allow one to deduce the TLS density close to the JJs, which could be compared for samples fabricated on different substrates.

The strongly coupled TLS ($g/2\pi > 1\,\mathrm{MHz}$) observed in this work (see App. Fig. D.2 b) most probably reside in the small Josephson junctions of the qubit. Their dipole moments are about ten times smaller than those of interface-TLS which may stem from molecules adsorbed at the circuit surface, that may possess larger dipole moments than tunneling ions inside the amorphous AlO_x tunnel barrier. Also, TLS hosted in the stray junctions show rather smaller dipole moments than interface-TLS. On the other hand, a similar tendency was observed when comparing junction-TLS with sample-TLS detected in 50 nm thick AlO_x, as reported in Fig. 4.3 (a). These observations suggest that generally, junction-TLS have smaller dipole moments than other TLS, which may be explained by screening of junction-TLS by the superconducting condensate whose evanescent wave function penetrates the thin junction's tunnel barrier. Possibly, examination of TLS in very thin AlO_x films using here-developed TLS-sensors would help verifying screening by Cooper-pairs.

The evaluation of TLS positions are the easiest in an Xmon-type qubit while it becomes complicated or maybe unfeasible using a floating transmon, due to stray fields induced between the qubit island and ground. A major upgrade of the setup used here would be to enlarge the global electrodes beyond the sample chip footprint, which would guarantee the field homogeneity all over the sample chip. Further, improved adjustment of the electrodes' distances is expected using the sample holder introduced in Fig. 3.10, where the chip is fixed directly to the box without a ceramic spacer. The bottom electrode can then be placed inside the cavity underneath the chip. Alternatively, the bottom electrode can be deposited on the chips' back side, which however would neglect the appealing portability of the setup.

Summarizing, a method has been developed and demonstrated to analyze the TLS positions at qubit interfaces. The underlying experimental setup is very simple and portable, allowing analysis of ready-made qubit samples while the method outcome directs efforts to improve qubit fabrication. Resolution of TLS positions for varying treatments of different interfaces before, during, and after fabrication can provide insight to the relevant process reducing TLS-related losses. The E-field simulations have shown that at film edges, the qubit field is concentrated at the substrate, where we observe practically all surface-TLS. To confirm the here-developed method, it would be intriguing to analyze TLS locations in samples with suspended film edges [Chu+16] where the qubit fields may be less focused, which would result in a broader TLS position distribution.

5 Problems And Solutions

5.1 Chlorine-based dry-etching of aluminum films

In the first part of this work, qubit and qubit-derived TLS-sensor samples were fabricated using the clean-room facilities of the Nanostructure Service Laboratory at KIT. An inductively coupled plasma (ICP) was used to pattern micro-structures into the 50 nm thick aluminum film (see Sec. C for film deposition details). Three main recipes were tested in this scope whose parameters are listed in Tab. 5.1.

#	Ar(sccm) Cl(sccm)	time (s)	P_{RF} (W)	P_{ICP} (W)	$C_{RF,1}$ (%) $C_{RF,2}$ (%)	$C_{ICP,1}$ (%) $C_{ICP,2}$ (%)
1	20 2	110	100	100	69 57	30 42
2	2 12	40	100	200	67 57	31 39
3	2 12	30	100	150	74 59	35 42

Table 5.1: Three main recipes tested in this work, used for dry-etching of aluminum films in the inductively coupled plasma (ICP) device. The last one lead to best-looking film edges shown in Fig. 5.1 (d) where the spurious holes along the edge are explained in Sec. 5.2. The power applied to the magnetic coil that generates the plasma is called P_{RF} while P_{ICP} is the power pumped into spacial plasma oscillations used to bombard the sample surface. The capacitances C_{RF} and C_{ICP} are start values used for automatic impedance matching to direct the applied power. The pressure inside the process chamber was 10 mTorr, the dc-bias 500 V, and the sample temperature 20 °C.

The first etching recipe (see #1 in Tab. 5.1) had a large physical and a small chemical component. It caused two spurious effects that hindered creation of a good galvanic contact to deposited films in the subsequent lithography step. The first effect was a metallic and up to 50 nm high fence along the etched film edge, visible in an SEM record in Fig. 5.1 (a), that we explain by sputtered and back-deposited aluminium. Second, the required long etching time lead to a well-known polymerization of the resist mask due to ion implantation. The resulting thin "skin" was observed to stick on top of the metal film due to capillary forces during stripping of the sample in a resting N-Eethyl-2-pyrrolidon (NEP) bath. Also stripping in an ultrasonic NEP bath did not change the outcome, as reported by colleagues.

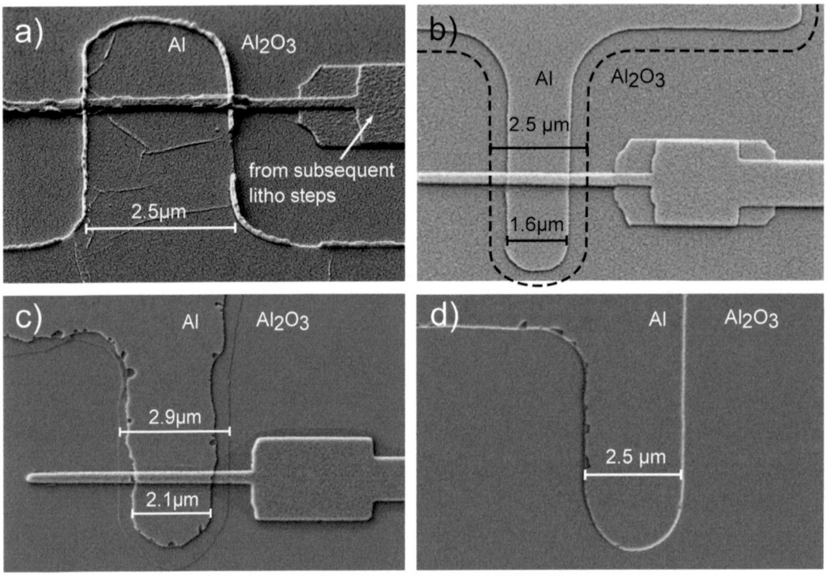

Figure 5.1: **a** The first etching recipe (see #1 in Tab. 5.1) had a large physical and a small chemical component. One spurious effect was a metallic fence around the film extent that we associate with back-deposition of sputtered aluminium. Second, the required long etching time lead to a well-known polymerization of the resist mask due to ion implantation. The resulting thin "skin" may stick on top of the metal film during stripping. **b** The second recipe (#2) employs a strongly chemical plasma. First, relatively long etch times were used to exploit the under-etching that saturates and creates a smooth film edge: In the underlying SEM picture, the dashed line denotes the resist edge. **c** After exchange of the ICP turbo pump, the under-etching (recipe #2) lead to wavy film edges. **d** A further decreased forward power and etching time in recipe #3 finally led to well-defined film edges. The "mouse bites" at film edges visible in (c) and (d) are explained in Sec. 5.2.

The second etching recipe (see #2 in Tab. 5.1) used a strongly chemical plasma that isotropically etched the film under the resist edge. In Fig. 5.1 (b), the initial resist edge is indicated by the dashed line. We exploited the saturation of this under-etching to obtain a smoothly formed film edge shown in (b). However, after exchange of the ICP turbo pump, this effect did not work properly and lead to wavy film edges: the sample shown in Fig. 5.1 (c) has been etched for an even longer time of 53 sec.

The third and final recipe #3 is deduced from the previous one by a slightly decreased forwarded power and etching time, leading to a well-defined film edge depicted in Fig. 5.1 (d). The "mouse bites" along the edges visible in (c) and (d) are due to aluminum corrosion caused by chlorine residuals in contact with moisture in the atmosphere. The sample shown in (b) did not show such hole even some months later. It has been covered with a thin gold film for better SEM resolution (this was not practiced with following samples), which possibly isolated the chlorine from moisture. Further details on Al corrosion are presented in the following section.

5.2 Void formation and corrosion in aluminum films on sapphire

It was observed that holes emerge in the Al film on sapphire wafers (see Sec. C for deposition details) roughly ten days after deposition when the wafer was stored at ambient conditions. Equally, annealing the wafer at 200 °C in atmosphere for longer than 10 min lead to the same effect. Film holes are unwanted since they may bury resist residuals and thus increase dielectric losses in the circuit. In Fig. 5.2, such holes are visible on the SEM picture of the patterned film via dry-etching, where the "mouse bites" along the film edges are an independent effect reported later in this section. Dutta et al. [Dut+12] reports this puzzling artifact proposing the origin of holes to stem from oxidation of aluminum at the substrate-film interface leading to formation of hollow hillocks that tear apart at a certain point. Hereby, oxygen is supposed to diffuse directly through the Al film or through fine film pores. This effect is enforced by tensile stress due to mismatch of thermal expansion between Al and Al_2O_3. The latter statement is supported by our observations that the largest voids do not emerge at distances below ~5 µm from film edges (see Fig. 5.2) where the film is less stressed. Film strips narrower than ~5 µm, for example, seem not to have voids at all, as visible in Fig. 5.1 (c) and (d). To trace the substantial difference between the voids and the "mouse bites", a following experiment was done: wafers with newly deposited and non-patterned Al films were baked at 200 °C for 10 min. Before annealing, the films were scratched with a scalpel. The voids appeared all over the film but not within the quoted distance around the scratches. Especially, no holes emerged at film edges.

While the work [Dut+12] does not propose clear solutions to avoid the film holes, we suppose that increasing the film thickness from 50 µm to, e.g., 100 µm will reduce the void density. Further, since voids do not appear close to film edges where dielectric losses of transmon qubits are the strongest, the void formation is most probably not influencing the qubit losses but rather the aesthetics.

Similar to the voids reported above, the "mouse bites" visible in Fig. 5.2 have been observed to emerge after some days of storage at ambient conditions or after wafer annealing. This is a well known and precisely reported problem [Hes82; Ric+16]. When etching Al with a chlorine-based plasma, chlorine residuals remain at the etched film edge. Once brought in contact with atmosphere humidity, hydrolysis of chlorine takes place leading to film corrosion. This effect may be enforced by carbon contamination and the physical damage to the Al surface. A simple deionized water or TMAH rinse are reported to strongly weaken the corrosion. Equally, oxygen or hydrogen plasma cleaning mostly removes the Cl residuals. A more sophisticated method is a post-etching fluorine-plasma treatment that substitutes chlorine in the Al-Cl groups which are then less aggressive to Al and remain water soluble. Due to time-restrictions the here-listed methods have not been applied in this work.

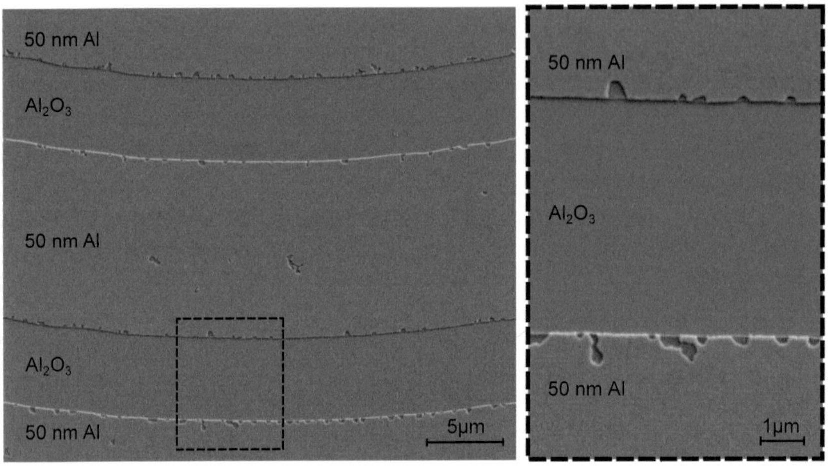

Figure 5.2: SEM picture of a chlorine dry-etched Al film on a sapphire wafer. Two effects are observable: formation of voids inside the film and emergence of "mouse bites" at film edges, both appearing after sample storage at ambient conditions for longer than about 10 days or annealing at 200 °C in atmosphere for more than 10 min. Film holes are unwanted since they may bury resist residuals and thus increase dielectric losses in the circuit. The void formation remains a puzzling problem reported in the work by Dutta et al. [Dut+12]. However, observations have shown that the voids emerge far away from film edges where dielectric losses of transmon qubits are dominant. Thus, the film holes should not increase qubit energy relaxation. The "mouse bites" are a well known and precisely reported problem [Hes82; Ric+16]. They result from Al corrosion due to chlorine residuals brought into contact with the atmosphere humidity. A simple water rinse strongly reduces this effect while fluorine-plasma treating is reported to fully substitute chlorine in Al-Cl compounds by fluorine, that are then less reactive and remain water-soluble. Due to time-restrictions, these improvements have not been applied in this work.

5.3 Optimization of the E-Beam lithography

As described in Sec. 3.2.2, the mask used for the Josephson junction (JJ) formation is written via an electron beam (E-beam) into a double-resist that covers the qubit island patterned in the previous lithography step. Hereby, the already existing metallic structures on the wafer may lead to deflections of the E-beam due to charging and direct scattering. This effect depends on the metallic structures geometry, as will be shown later in this section. Since test-JJs, which are useful for fast sample characterization, have to be comparable with the qubit-JJs, the probe-pads for test-JJs should emulate the environment (within $\sim$100 µm distance) of the qubit-junctions. The resulting test-JJ probe-pads are shown in Fig. 5.3 (e) while the qubit geometry is shown in Fig. 3.4 (a). In the following, the empiric derivation of such probe-pads is described.

First test-junctions were fabricated using floating probe-pads shown in Fig. 5.3 (a), where the red areas denote the aluminum film. The mask design shown in the inset of (b), and used for double-angle junction formation was based on the bridge-free technique [Lec+11] while the table in (c) contains the scaling factors of the E-beam-deposited charge density of each design layer. The junction pads used for galvanic contact to the underlying film had two different shapes: wide and narrow rectangles with a same area, as shown in Fig. 5.3 (b). Hereby, the narrow pads serve for tests only.

In Fig. 5.3 (d), SEM-pictures of the resist mask ("mask") after junction formation, and the junction shape ("film") after resist stripping are shown. The junctions having wide pads are overexposed. This is indicated in the "mask" photograph by the destroyed top resist areas that are supposed to be suspended, and in the "film" picture by the double wires, where the red and blue areas denote the shape of wanted first and second deposited layers, respectively. On the other side, the junctions with narrow pads are well-shaped. This indicates that the mask suffers from proximity effects which may be enforced by the floating probe-pads that are charged by the scanning E-beam and deflect the flying electrons towards the junction center.

Since the qubit-junctions have a different metallic environment compared to the floating probe-pads, the next version of probe-pads was chosen to emulate the same environment in order to make test-junction comparable to qubit-junctions. Figure 5.3 (e), (f) and (i) contain respectively the resulting probe-pad where the green area symbolizes the Al film, the unchanged junction design with wide pads, and the charge density table almost unchanged compared to the previous lithography. The resulting junction shown in the bottom SEM picture of (f) indicates the contrary result to the previous tests: underexposure of the bottom resist. This shows the substantial difference between floating probe-pads and the pads surrounded by a ground plane. After some further design improvements shown in (g) and (h), bridge-free and Dolan-technique based Josephson junctions were reproducibly fabricated, respectively.

5.4 Small fabrication problems

5.4.1 Thermal deposition of Nb alignment marks

As shown in Fig. 3.3, a 70 nm thick Niobium (Nb) film thermally deposited in the PLASSYS, forms crosses for the EBL alignment. Hereby, the lift-off mask (S1805 photo-resist) is the same used for dry etching of the main structures into the Al ground plane. When thermally depositing Nb in the PLASSYS the heat radiation from the crucible can severely damage the resist or cause polymerization which can hinder the following stripping. The S1805 resist turned out to roll up at corner protrusions, as shown in the photograph in Fig. 5.4 a) depicting a niobium cross with shaded inner corners explained in (b). One possible solution is to reduce the Nb deposition rate from 0.2 nm/s to 0.1 nm/s and to include a 5 min break in the middle of the deposition process to let thermalize the heat stored in the mask. Herewith, the rolling up of the cross corners occurred only for one of 32 crosses and close to the end of the deposition process so that the unwanted Nb film was thin enough to be ignorable during the EB alignment. Since the Nb crosses are deposited after etching the Al ground plane, they stick directly to the sapphire wafer and seem stable and reusable during the alignment with the 50 keV EB writer, which was not the case when using Nb crosses deposited on top of the Al ground plane.

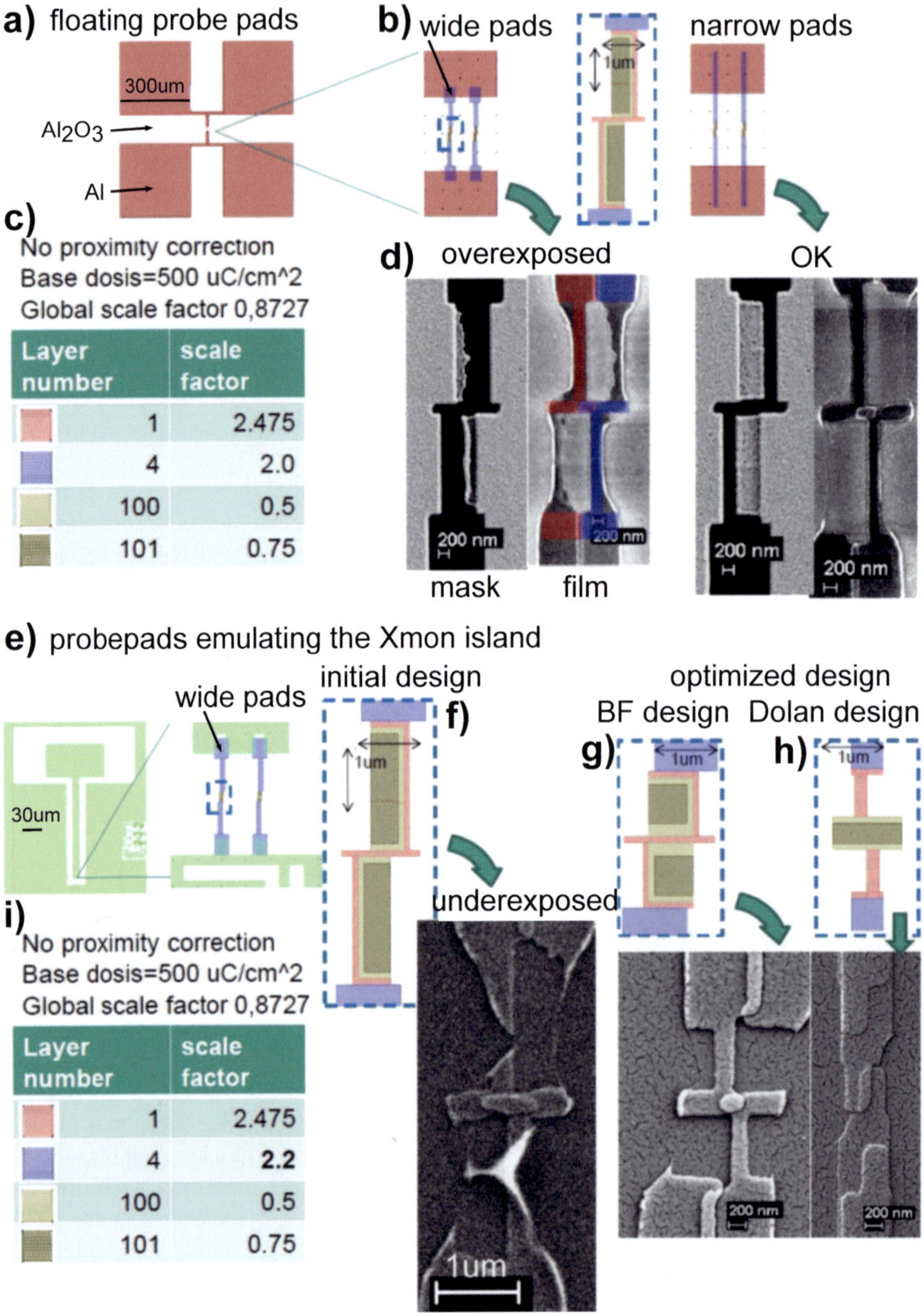

Figure 5.3: During optimization of the Josephson junction design, a main observation was that the E-beam lithography is very sensitive to already-existing metallic structures on the wafer. For this reason, it seems practical if the environment of the qubit-junctions and the test-junctions is identical to have both junction types comparable. Further details are given in the text.

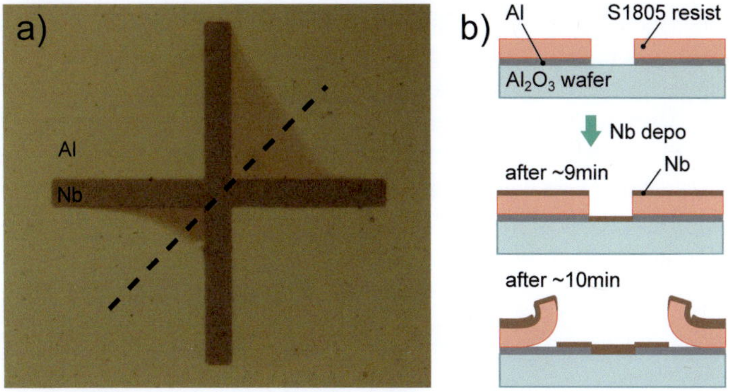

Figure 5.4: **a** A photograph of a bad 70 nm thick Nb cross supposed for EBL alignment, and deposited in the PLASSYS. Due to heat radiation from the Nb crucible the S1805 optical resist rolls up at the corner protrusions leading to the shadow-like fade-out in the inner cross corners. **b** A sketch showing the wafer cross section (dashed black line in **a**) before Nb deposition, ca. 9 min after start of deposition and after ca. 10 min when the resist starts to roll up. The solution applied in this work was to reduce the Nb deposition rate from 0.2 nm/s to 0.1 nm/s and to introduce a break of 5 min in the middle of the deposition process to let the resist cool down. The dots on the Al film shown in **a** are voids described in the section. 5.2.

5.4.2 Preventing ESD when removing the microchip from the dicing foil

It is a known problem that Josephson junctions of qubit samples with large electrodes can be destroyed by electrostatic discharge when pealing off the microchips from the dicing foil. For this reason conducting and UV-treatable dicing foils are available, but still the risk of ESD remains. A method was developed together with Lucas Radtke to release the microchips from the UHP-0805M6E2 dicing foil. After dicing, a piece of the foil with the wafer with two wide margins is cut out, as depicted in Fig. 5.5 (a). The wafer with the foil is placed into an N-Eethyl-2-pyrrolidon (NEP) bath at room temperature, as illustrated in Fig. 5.5 b) and put on a 50 °C warm hotplate. After 10 min, the chips start to fall off the foil. They are then put into another NEP bath at 90 °C for 6 h to properly remove the protecting photo-resist used for dicing. The temperature of 50 °C was chosen since one can shore up the eventually

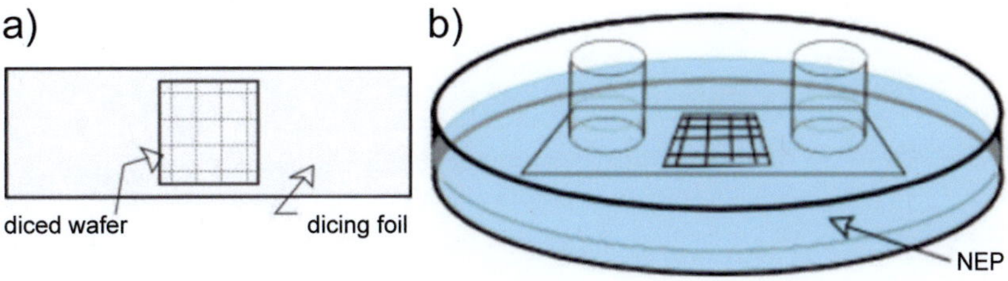

Figure 5.5: A sketch (by L. Radtke) describing a technique to release the diced wafers from the dicing foil UHP-0805M6E2 with a minimum risk of electrostatic discharge. **a** After dicing: cut out the peace of the dicing foil with the wafer. **b** Dip the foil peace into an N-Eethyl-2-pyrrolidon (NEP) bath and place it on a 50 °C warm hotplate. After 10 min the chips start to slide off the foil.

shaking hand onto the hotplate while fishing out the chips with tweezers, which peel off the foil. This method was tested with wafers whose both sides were covered with protecting resist, expecting faster foil removal. Interestingly, the chips did not fall off the foil in a cold NEP bath after $\sim 10\,\text{min}$. Thus, NEP seems to diffuse directly through the foil (the diffusion is slowed down at room temperatures) rather than between the wafer and the foil, which is supported by the fact that two inch wafers need the same time to peel off in the 50 °C warm NEP bath, as reported from colleagues. Not necessary to mention that this recipe may work completely different for other dicing tapes.

6 Conclusions And Outlook

Two-level-systems (TLS) are parasitic GHz resonances emerging in cold solids, especially in the circuit materials of superconducting qubits. They possibly constitute the main impediment to development of an integrated superconducting quantum processor, which makes TLS research very important and exciting. The experiments of this work are twofold. On the one side, the first TLS-sensor derived from an Xmon qubit architecture was developed, fabricated and tested. This tool enables one to sense TLS in a broad choice of dielectrics to learn more about the TLS' microscopic nature. On the other side, non-invasive methods to investigate spurious TLS in ready-made qubit samples have been tested, which equally provide information about TLS' origin and their possible locations in the quantum circuit.

The here-reported experiments were performed at low temperatures (30 mK) using a cryogenic refrigerator whose microwave setup was upgraded in the course of this work to allow for dispersive readout of the qubit quantum state. Further, a sample holder was developed and optimized to operate quantum circuits in an electromagnetically unperturbed environment while allowing for TLS-tuning via two degrees of freedom: elastic strain transferred to the sample chip using a piezo actuator, and static electric field from global electrodes used in the second part of this work.

TLS-sensors

The here-elaborated TLS-sensors are based on a qubit shunted with a small sample capacitor hosting a dielectric of interest. Due to the small electrode distance of the sample capacitor, strong electric fields are induced by the qubit, which couples the latter to TLS hosted inside the sample dielectric. The first TLS-sensor derived from an Xmon qubit architecture was fabricated using the clean-room facilities of the Nanostructure Service Laboratory at KIT. The characterized samples serve as a proof of principle that such sensor architecture is suitable for investigation of sample-TLS. Further, this architecture enables application of electric fields via on-chip electrodes, and elastic strain, to tune the probed TLS and study their properties for different internal TLS parameters. Especially, this sensor allows for discrimination of TLS residing in the sample dielectric from the intrinsic ones naturally hosted in the qubit circuit. This basic sensor property as well as the device tunability within $\sim$1 GHz were not given in previously used architectures based on microwave resonators.

The first tested sensor version showed a rather poor TLS resolution due to a short energy relaxation time which resulted from the over-sized sample capacitor. The next generation sensor has been fabricated but yet not experimentally characterized. It is coupled to a smaller shunt capacitor that was fabricated using electron-beam lithography. Another alternative sample capacitor design is a coplanar nano-gap designed and fabricated in this work. It allows for probing of TLS residing in thin and thick films, as well as on the electrode surface where decoherence-bringing TLS are assumed in the state-of-the-art qubits.

The first sensor version allowed us to determine the sample dielectric loss tangent in good agreement with measurements in comparable setups using microwave resonators. Further, it allowed us to measure the components of TLS electric dipole moments that are parallel to the qubit field inside the sample dielectric. Due to poor statistics, we only carefully state that the results indicate that dipole moments of TLS hosted in the 50 nm thick AlO_x are about three times larger than those of TLS residing in the tunnel barrier of the qubit Josephson junction. This points to a fundamental difference between TLS in dielectric films and those in very thin films that rather resemble an interface: A question to be further investigated in the following generations of TLS-sensors.

TLS location and classification in ready-made qubits
Intrinsic TLS in Xmon qubit samples have been examined. From tunability characteristics of detectable TLS by strain transferred via a piezo actuator, and by dc-electric fields generated by two global electrodes, conclusions have been made about possible TLS locations and their nature. Besides TLS that are strongly coupled to the qubit, two classes of weakly coupled TLS could be discriminated from these experiments, as summarized in table 6.1.

The first and well-known class are TLS hosted in the small Josephson junction, which couple the strongest to the qubit but appear rarely in contemporary devices due to the junction size. Like all here-detected TLS, they couple to elastic strain but not to the global electrodes since the inside of the junction tunnel barrier is free of dc-fields due to the transmon regime of the qubit. The TLS of the second class exhibit equal tunability characteristics and reside most probably in the large-area stray junctions (connected in series with the small qubit junction) where they experience only a tiny fraction of the oscillating qubit field and correspondingly weakly couple to the device. On average we have observed 15 such TLS per GHz. The third class with a comparable density are so-called interface-TLS which are dc-field tunable, and predominantly reside at the substrate-vacuum interface and on top of the natively passivated aluminum film of the qubit.

#	TLS location	density (1/GHz)	qubit coupling $g/2\pi$ (MHz)	dipole moment (eÅ)
I	small junction	<1	2 to 10	0.02 to 0.2
II	stray junction	15	0.04 to 0.4	0.1 to 1.4
III	predominantly at surfaces	23	0.01 to 0.3	0.1 to 2

Table 6.1: Summary of three distinguishable classes of TLS detected in the Xmon samples from UCSB, USA.

The location of first and second classes of TLS is supported by simulation results which show that except of the tunnel barrier of the Josephson junctions, no other spot in the qubit circuit exists where the qubit field is strong enough to couple to TLS while the dc-field from global electrodes vanishes. Further, it is difficult to imagine a mechanism that allows a charged TLS residing outside of the junction to couple to the qubit only: e.g., attempts to identify the 2nd class as magnetic TLS remain unconvincing. Regarding the measured TLS densities we state that stray junctions essentially contribute to dielectric losses, and need to be shorted [Dun+17] or omitted by fabricating junctions in two lithography steps [Wu+17].

Interface-TLS (3rd class) have been shown to reside predominantly at the sample surface. Only few detectable TLS are hosted at the interface of the sapphire substrate and the epitaxially grown aluminum film as expected in the report by Martinis et al. [MM14]. Further, while it is likely that TLS exist in the native oxide covering the qubit film, the electric field strength inside the dielectric is reduced by about a factor of ten so that a qubit energy relaxation time longer than 100 μs would be required to detect them. Interestingly, this number is the limit experimentally achieved with transmon qubits, which may indicate that the most coherent transmons suffer from dielectric losses induced in the native aluminum surface oxide.

The analysis of TLS locations proposes that most detected interface-TLS are residing within $\sim$50 nm from the film edge. This can be expected since the electric field induced by the qubit is concentrated in this region. An intriguing experiment to test the here-developed technique would be to analyze TLS positions in samples with suspended film edges [Chu+16] which may exhibit a more diluted qubit field, but on the other side offer larger surface for TLS-forming adsorbates. In another future experiment, one could place local electrodes close to the qubit to tune TLS residing nearby. The contrast to the global electrode that addresses all interface-TLS could help resolving TLS locations along the qubit film edge. For example, one could verify whether TLS preferably emerge close to the Josephson junctions in Xmons fabricated on silicon wafers [Dun+17], and compare the findings with samples on sapphire.

7 Appendix

A Capacitance estimation of the dc-wiring

The capacitance $C_{\mathrm{tg/bg}}$ of the top/bottom electrode to ground is dominated by two things. The RC-filters at the 1 K stage have a capacitance to ground of about 200 pF. The dc-wiring in the cryostat consists of constantan wires (radius $r \approx 0.065$ mm) that lead from the 4 K plate down to the mixing chamber. Hereby, at every temperature stage, they are wound around a thermalization spool and glued with GE-varnish. The capacitance to ground of one wire is almost fully determined by the spools. Thus, it can be calculated from the capacitance per unit length C_{l} of a wire placed at a distance h (from its center) above an infinite metallic plate:

$$C_{\mathrm{l}} = \frac{2\pi\varepsilon_0\varepsilon_r}{\ln\left(\frac{h}{r} + \sqrt{\left(\frac{h}{r}\right)^2 - 1}\right)}, \tag{7.1}$$

where $\varepsilon_r \approx 4$ is the varnish relative permittivity at low temperatures. About 1 m of constantan wire is wound around the spools. Estimating $h \approx 2r$, the wire capacitance to ground and herewith the capacitance $C_{\mathrm{tg/bg}}$ is about $200\,\mathrm{pF} + C_{\mathrm{l}} \cdot 1\,\mathrm{m} \approx 400\,\mathrm{pF}$. This is a rough estimation, which however shows that $C_{\mathrm{tg/bg}}$ is by at least three orders of magnitude larger than all other capacitances shown in the circuit diagram in Fig. 2.6 (b), so that the assumption of equal voltage drop $V_{\mathrm{tb}}/2$ across C_{tg} and C_{bg} holds.

B Basic fabrication recipes

B.1 Plasma cleaning recipes

The descum recipe is used to remove resist residuals in the developed areas of the mask, see Tab. B.1. The Argon plasma is used to remove the native oxide on top of the Al film to create galvanic connections to subsequently deposited metallic films, see Tab. B.2.

B.2 Ti gettering

The gettering process is applied to improve the vacuum in the load-lock of the PLASSYS shadow evaporation tool. About 12 nm Ti are deposited at a deposition rate of 0.1 nm/s while

Descum (Ar/O_2: 5 sccm/10sccm)				
Cathode	Discharge	Beam	acceleration	neutralizer
7 V	40 V	200 V	110 V	10 V (emission)
5 A	0.1 A	10 mA	0.9 mA	8 V
				10 A

Table B.1: Parameters of the KSC 1202 power supply unit that controls the KDC 40 ion source to generate the remote Oxygen-Argon plasma used to descum or clean samples in the PLASSYS MEB550s shadow evaporation tool. Typical application time is 2 min. KDC 40 and KSC 1202 are purchased from Kaufman&Robinson, Inc. (www.ionsources.com). The load lock pressure is ~ 1 to 5×10^{-3} mbar.

Argon ion-milling (5 sccm Ar)				
Cathode	Discharge	Beam	acceleration	neutralizer
7 V	40 V	400 V	90 V	15 V (emission)
5 A	0.15 A	15 mA	0.9 mA	8 V
				9 A

Table B.2: Parameters for Oxygen-Argon cleaning plasma used in the PLASSYS MEB550s shadow evaporation tool. Typical application time is 2 min. The remote plasma is generated with a KDC 40 ion source that is controlled by a KSC 1202 power supply unit, both purchased from Kaufman&Robinson, Inc. (www.ionsources.com). The load lock pressure is $\sim 1 \times 10^{-3}$ mbar.

the sample shutter is closed. The static pressure of the load lock (connected to the process chamber) falls from $\sim 8 \times 10^{-8}$mbar to $\sim 3 \times 10^{-8}$mbar after gettering.

C Fabrication parameters for the TLS-sensor

Experiments with samples fabricated on wafer L3 are reported in Sec. (4.1).

Wafer preparation and Al film deposition

3 inch Sapphire wafers (3″/C-Plane/500±20/DSP, from SIEGERT WAFER GmbH, Charlottenburger Allee 7, 52068 Aachen, GERMANY) were prepared as follows.

1. Clean with piranha solution (mix of sulfuric acid H_2SO_4 and hydrogen peroxide (H_2O_2)) to remove organic residuals.
2. Load the wafer into the PLASSYS shadow evaporator and pump the load lock for 2 h at 200 °C. Cool down to room temp. over night.
3. Apply oxygen-argon cleaning recipe (Tab. B.1) for 1 min.
4. Ti gettering
5. Deposition of SI50nm Al that will define the qubit ground plane (room temperature, deposition rate of 1 nm/s).

6. Spin-coating the wafer with protecting S1818 photo-resist (~2 µm thick) and subsequent dicing into seven 20 mm × 20 mm-large wafers.

7. Remove the protecting resist:

   a) 15min in NEP (N-Ethyl-2-pyrrolidon) at 90 °C, then 1 min in an ultrasonic bath at low power

   b) 1 min in Isopropanol or 2-propanol bath in an ultrasonic at low power

   c) second bath in Isopropanol or 2-propanol, 1 min in an ultrasonic at low power

   d) bi-distilled water, 1 min in an ultrasonic at low power

   e) blow dry with nitrogen gun

Optical lithography

In the first optical lithography step, the main structures are patterned into the ground plane, and Nb alignment marks are deposited for the electron-beam (EB) lithography by subsequent steps listed in Tab. C.1. Details of the EB mark deposition are shown in Fig. 3.3. The hard mask used to expose the photo-resist S1805 is formed from chrome on a soda substrate, and is labeled M0042. Alignment and exposure was performed in a mask aligner Carl Suess MA6 (Xe 500W lamp, wave lengths: 240, 365, 405 nm). Etching was done with inductively coupled Ar-Cl plasma (ICP, Oxford Plasma Technology, Plasmalab 100 ICP180) while the wafers were placed on a 5 inch Si carrier wafer. After deposition of alignment marks, the wafer was striped in NEP (1h, 90°).

Electron-beam lithography 1

In the first E-beam lithography step, the qubit Josephson junctions (JJs) are fabricated (see Fig. 3.4 for designs) by following steps:

1. 2018Mar23: Spin-coat the wafer with a resist stack (A-4 PMMA on top of El-13 copolymer, see Tab. C.2) and cover it with a conducting gold layer.

2. 2018Mar26: E-beam exposure (JEOL JBX-5500ZD, 50 keV acceleration voltage).

3. Dip the wafer for 10 s in a Lugol solution (15%) to remove gold.

4. Develop the mask in a H_2O/2-propanol (1:3) solution at 6 °C for 90 s. Wash in bidest. water and blow dry.

5. Load into PLASSYS shadow evaporator:

   a) Pump for ~2 h → static pressure 2.1×10^{-7}mbar.

   b) Apply descum (see Tab. B.1) for 2 min.

   c) Apply Ar ion milling (see Tab. B.2) for 2.5 min.

   d) Apply Ti gettering.

   e) Deposit 30 nm Al (angle 24°, deposition rate of 1 nm/s).

Resist application (photo-resist S1805)							
date (2018)	accel. (rpm/s)	v_{ramp} (rpm)	t_{ramp} (s)	v_{main} (rpm)	t_{main} (s)	T_{hotplate} °C	t_{hotplate} (s)
Mar23	700	500	1	6k	60	115	60

Optical lithography (Mask aligner, hard exposure, intens. 13 mW/cm^2)				
date (2018)	expos. time (s)	alignment gap (µm)	developer/ H_2O (1/1)	development time (s)
Mar23	1.5	50	AZ-developer	30

Dry-etching (ICP, pressure 10 mTorr, dc-bias 500 V)						
date (2018)	etch time (s)	Ar/Cl (sccm/sccm)	P_{RF} (W)	C_1/C_2 (%/%)	P_{ICP} (W)	C_1/C_2 (%/%)
Mar23	53	2/12	100	71/57	200	30/42

Alignment mark deposition (PLASSYS, zero angle)				
date (2018)	metal	deposition rate (nm/s)	thickness (nm) (see Sec.5.4.1)	descum (min) (see Tab. B.1)
Mar23	Nb	0.1	70	2

Table C.1: Optical lithography and subsequent steps.

f) Static oxidation at 20 mbar for 4 min, 50 s.

g) Deposit 60 nm Al (angle –24°, deposition rate of 1 nm/s).

6. Lift-off in NEP (1 h, 90 °C).

7. Probe the normal resistance of test dc-SQUIDs on the wafer.

8. Bake wafer at 200° for 5 min to anneal the Josephson junctions.

9. Probe the normal resistance of test dc-SQUIDs on the wafer, the normal resistance typically increases by $4 \pm 8\%$.

Resist application (E-beam lithography)					
resist	acceleration (rpm/s)	v_{main} (rpm)	t_{main} (s)	T_{hotplate} °C	t_{hotplate} (min)
EL-13	1000	2000	100	200	5
A-4	1000	2000	100	200	5

Table C.2: Resist application parameters for the E-beam lithography.

Electron-beam lithography 2

The overlap junctions were formed after the second E-beam lithography (see Fig. 3.5 for designs):

1. 2018Mar27: Spin-coat the wafer with a resist stack (A-4 PMMA on top of El-13 copolymer) and cover it with a conducting gold layer (see Tab. C.2).
2. 2018Mar28: E-beam exposure in the JEOL writer (see Tab. 3.3 for doses).
3. Dip the wafer for 10 s in a Lugol solution (15%) to remove gold.
4. Develop the mask in a H_2O/2-propanol (1:3) solution at 6 °C for 90 s.
5. Load into PLASSYS shadow evaporator:
   a) Pump for $\sim$2 h $\rightarrow$ static pressure 2.7×10^{-7} mbar.
   b) Apply Ar ion milling (see B.2) for 2.5 min.
   c) Apply Ti gettering.
   d) Set Al deposition rate of 0.1 nm/s (angle 0°) in an oxygen atmosphere (oxygen flow 5 sccm, chamber pressure $\sim 6 \times 10^{-4}$ mbar) and wait for 2 min at closed sample shutter. Open shutter and deposit 50 nm Al.
   e) Deposit 100 nm Al (angle 0°, deposition rate 1 nm/s).
6. Lift-off in NEP (1 h, 90 °C).

Wafer dicing
The wafer was spin-coated with protecting S1818 photo-resist ($\sim$2 µm thick) and diced into six 6 mm × 6 mm-large wafers. The chips were removed from the dicing foil in an NEP bath as described in Sec. 5.4.2.

D Spectroscopy with applied elastic and electric fields: supplemental material

The angle α_{tb} between the dc-electric fields generated by the top and bottom electrodes is plotted in Fig. D.1. Statistics on strongly coupled junction-TLS observed in several cool downs are presented in Fig. D.2. Figures D.3 and D.4 contain raw data of recorded TLS traces under application of dc-electric and strain fields.

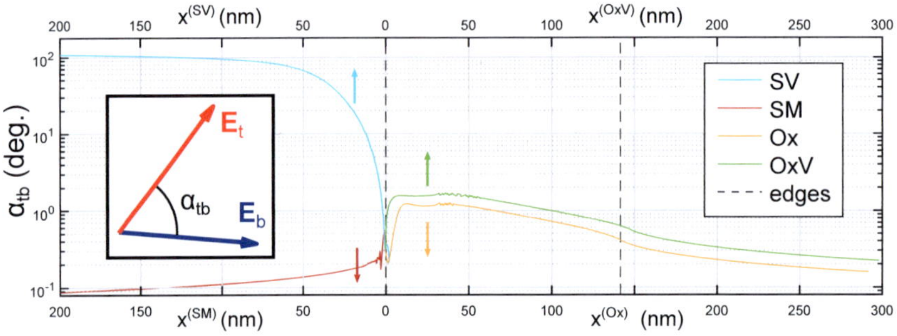

Figure D.1: Plot of the angle α_{tb} between the fields $\boldsymbol{E}_{t/b}$ (see inset) generated by the top/bottom electrodes. $x^{(i)}$ is the spatial axis at every interface while $x = 0$ denotes the film bottom edge, where all four interfaces intersect. We recognize that at film interfaces (SM, Ox and OxV interfaces), the mutual angle is below 2°, which means that $\boldsymbol{E}_t$ and $\boldsymbol{E}_b$ are practically parallel. This is a key point for the analysis of TLS locations at film interfaces. However, at the SV interface the angle increases fast with x so that a more involved analysis is required. Both methods are introduced in Sec. (2.3.5) and reported in Sec. (4.2.5).

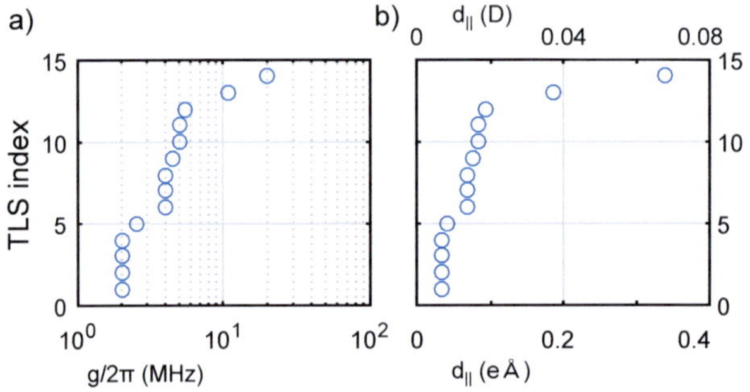

Figure D.2: **a** Coupling strengths g of TLS to the qubit, extracted from level splittings observed in the qubit spectrum. These well-known and so-called strongly coupled TLS are presumably hosted in the tunnel barrier of the qubit' small Josephson junction (JJ) [Mar+05]. The data shown here has been recorded in four independent refrigerator cool downs, with Xmon samples from UCSB, USA. **b** Minimum TLS electric dipole moments (assuming a symmetric TLS) deduced from g, the tunnel barrier thickness ~2 nm and the root mean square of the qubit field 5 μV induced inside of the JJ dielectric.

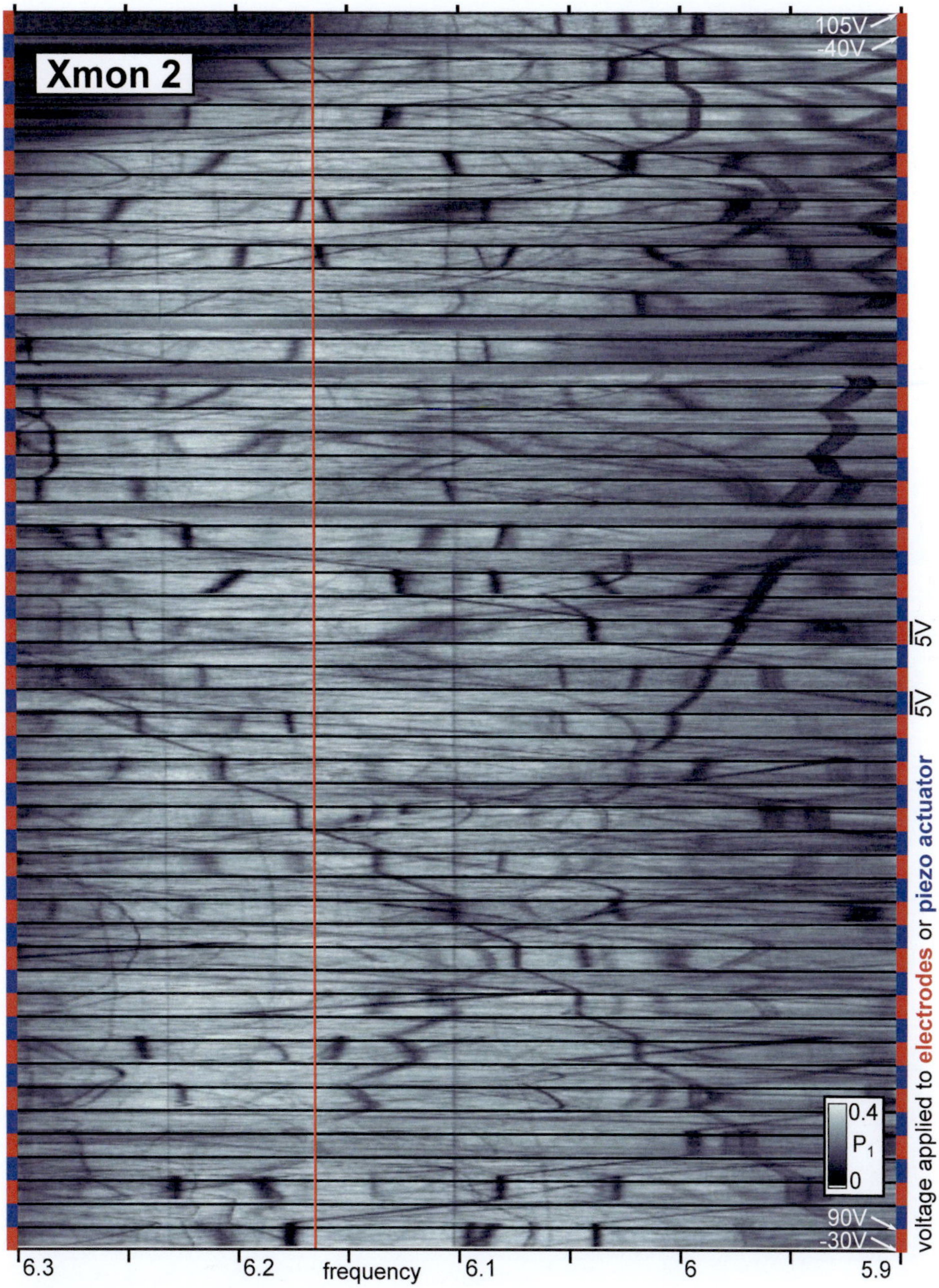

Figure D.3: Raw data recorded with Xmon 2 using the swap-spectroscopy protocol ($\tau = 5\,\mu s$). The red line ($\sim 6.17\,GHz$) is the qubit frequency at zero applied magnetic flux. The voltage interval of each data subset (red and blue margined data sections) is chosen to be narrow relative to the total ramped voltage range. This increases the probability that each detected TLS appears in at least two data sections, which is necessary to identify junction-TLS.

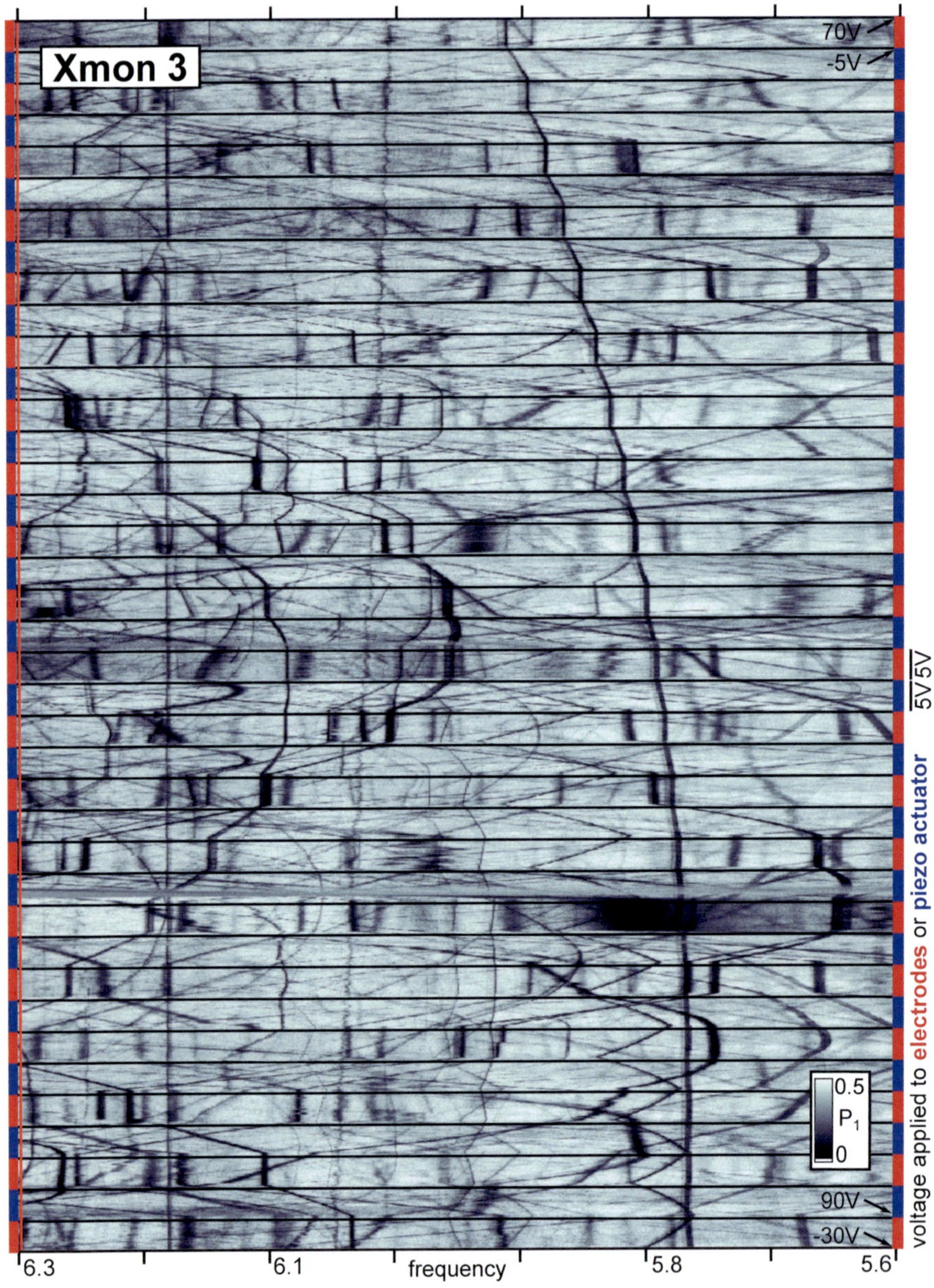

Figure D.4: Raw data recorded with Xmon 3 using the swap-spectroscopy protocol ($\tau = 5\,\mu s$). The red line ($\sim 6.3\,GHz$) is the qubit frequency at zero applied magnetic flux.

E Coupling strength to surface spins

Here, we roughly estimate the coupling strength g_S between the qubit and a magnetic spin $1/2$ residing on the surface of the leads connecting the qubit island to ground via two Josephson junctions connected in parallel. The coupling strength is defined as $g_S = g\mu_B B^{\mathrm{rms}}$, where the g-factor is assumed to be 2. The root mean square magnetic field B^{rms} is induced by rms vacuum fluctuations of the current $I^{\mathrm{rms}} = \sqrt{\mathrm{h} f_{01}/2L_{\mathrm{JJ}}}$ that flows through both leads. In the last formula, we neglected the geometrical inductance of the leads and regard only the Josephson inductance L_{JJ} estimated as follows. From the measured resonance frequency of the qubit $f_{01} = (\sqrt{8E_C E_J} - E_C)/\mathrm{h} \approx 6.4\,\mathrm{GHz}$ and the anharmonicity $\alpha/2\pi \approx 220\,\mathrm{MHz}$, we estimate the Josephson energy to be $E_J \approx \mathrm{h}\,25\,\mathrm{GHz}$, where the relation $\hbar\alpha \simeq -E_C$ was used valid in the transmon regime. Thus, the critical current $I_c \approx 25\,\mathrm{nA}$ per junction is obtained, using the relation $E_J = \hbar I_c/(2e)$. The Josephson inductance is hence $L_{\mathrm{JJ}} = \hbar/(2eI_c) \approx 6.6\,\mathrm{nH}$, and $I^{\mathrm{rms}} \approx 18\,\mathrm{nA}$. As simulated and reported in the PhD thesis by Audrey Bienfait [Bie16; Bie+17], the maximum magnetic field of ~ 14 nT is induced at film edges of a lead with a cross-section of $5\,\mu\mathrm{m} \times 50\,\mathrm{nm}$ carrying a current of 48 nA. Scaled down to I^{rms}, the magnetic field is thus ~ 2.6 nT. The surface spin thus couples by roughly $g_S \approx 40\,\mathrm{Hz}$.

F Deducing TLS positions with two dc-electrodes: supplemental material

Figure F.1 contains factors $\gamma_{t/b}$ recorded with Xmon samples 1 and 3. Figures F.2 to F.4 contain raw data of recorded TLS traces under application of dc-electric fields. Some comparisons of deduced dipole moment components are drawn in Fig. F.5, and in Fig. F.6, the inhomogeneity of applied dc-electric field at the location of Xmon 3 is shown. Further, Fig. F.7 illustrates the outcome of the TLS location analysis under small variations of distances between the qubit chip and the global electrodes. in figure F.8, the role of the dipole moment cutoff is discussed, used in the location analysis.

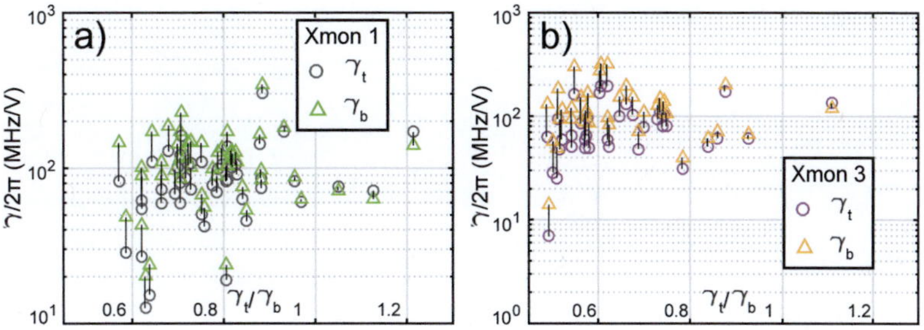

Figure F.1: a, b Fit factors γ_t and γ_b deduced from data shown in Fig. F.2 and Fig. F.2, and plotted vs. γ_t/γ_b. The vertical black lines interconnect γ_t (circle) and γ_b (triangle) of the same TLS for better visibility.

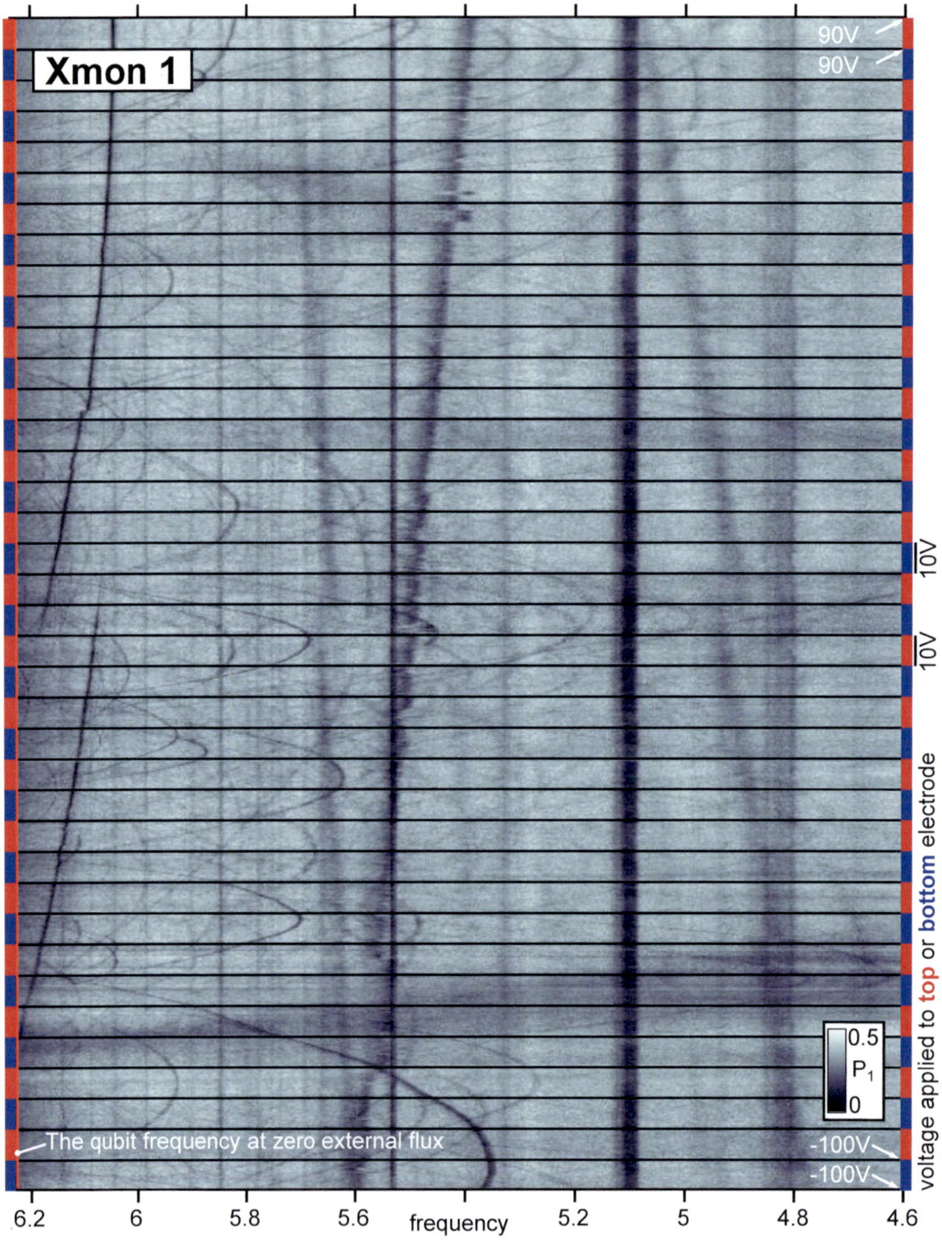

Figure F.2: Raw data recorded with Xmon 1 using the swap-spectroscopy protocol ($\tau = 1\,\mu s$), employed in the analysis of interface-TLS positions at the film edge. The horizontal dark traces stem from junction-TLS and are disregarded in this analysis. See following figures for more general properties of such data records.

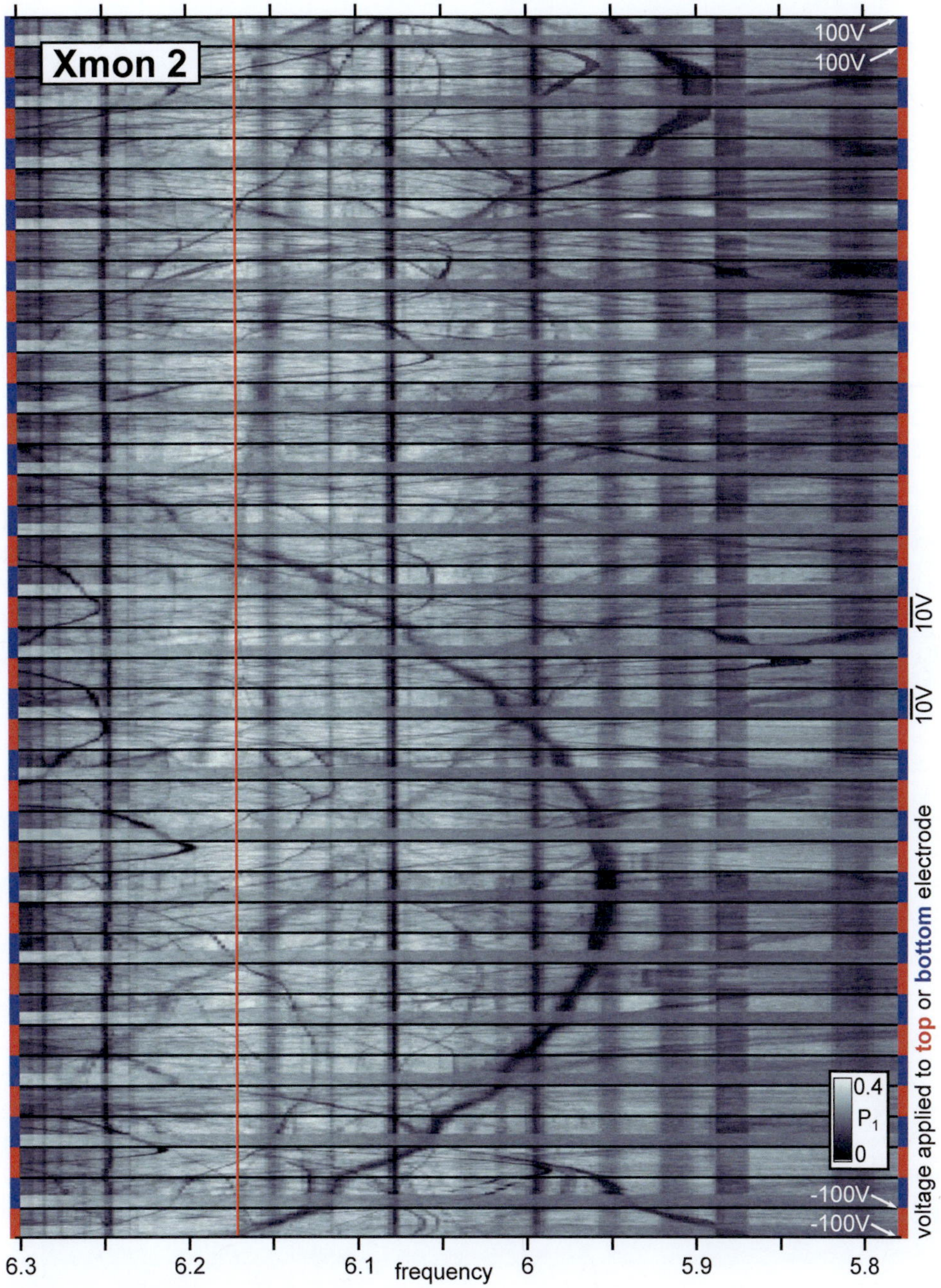

Figure F.3: Raw data recorded with Xmon 2 using the swap-spectroscopy protocol ($\tau = 6\,\mu s$). The voltage interval of each data subset (red and blue margined data sections) is empirically chosen to be narrow relative to the total ramped voltage range. This increases the probability that each detected TLS appears in at least two data sections, which is necessary to apply the analysis of TLS-positions.

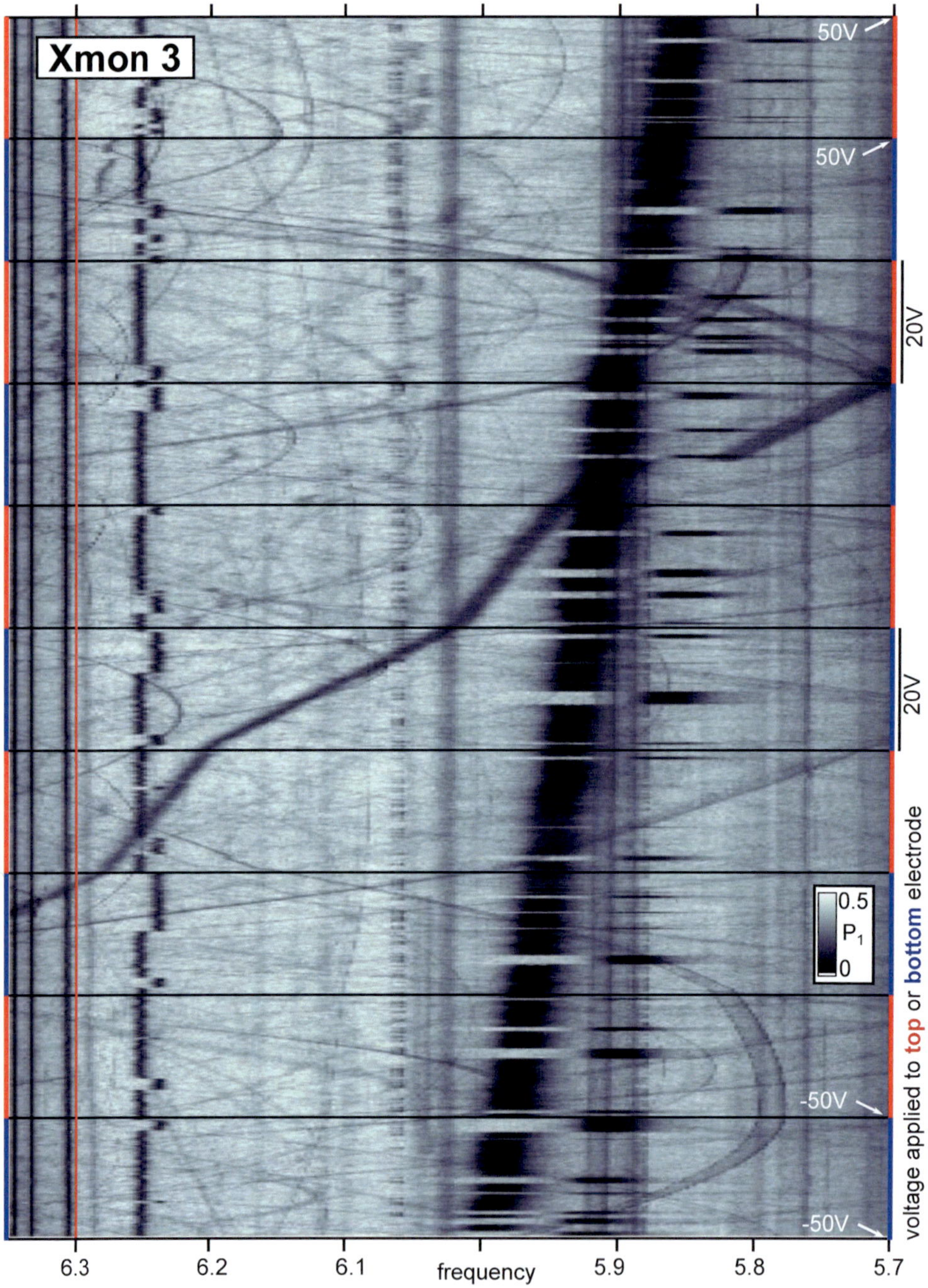

Figure F.4: Raw data recorded with Xmon 3 using the swap-spectroscopy protocol ($\tau = 5\,\mu s$).

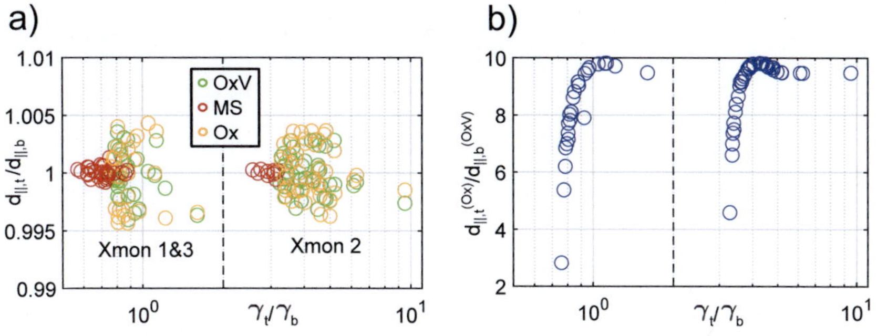

Figure F.5: **a** Plot of the ratio $d_{\|,t}/d_{\|,b}$ at each of the film interfaces (OxV, SM, Ox). Hereby, $d_{\|,t/b}$ is the TLS electric dipole moment component parallel to the dc-electric field $\boldsymbol{E}_{t/b}$ generated by the top/bottom electrode, respectively. The deviation is below 0.5 %, which quantitatively verifies the assumption that at film interfaces $\boldsymbol{E}_t$ and $\boldsymbol{E}_t$ are parallel. **b** Ratio of $d_{\|,t}^{(Ox)}$ derived from TLS locations at the Ox interface, and $d_{\|,t}^{OxV}$ derived from positions at the OxV-interface. This ratio equals the ratio $E_t^{(OxV)}/E_t^{(Ox)}$ of the fields at the OxV, and Ox interfaces, respectively. As expected, $E_t^{(OxV)}$ is about ten times higher than $E_t^{(Ox)}$.

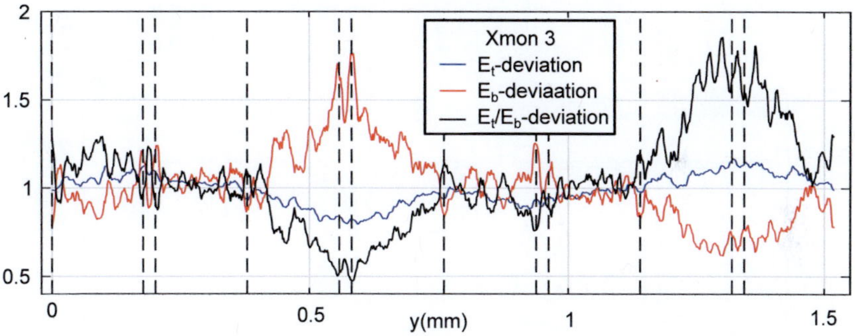

Figure F.6: Simulated variations of electric fields $E_{t/b}(y)$ induced by the top/bottom electrodes. The resulting variation of the $E_t(y)/E_b(y)$ ratio discussed in the main text and shown in Fig. 4.6.

G Source code, measurement and simulation data

Data evaluation was performed in Matlab. The functions used for the analysis of TLS locations, and for data plots in this work, as well as measurement and simulation data are published in the KITopen repository [Bil19].

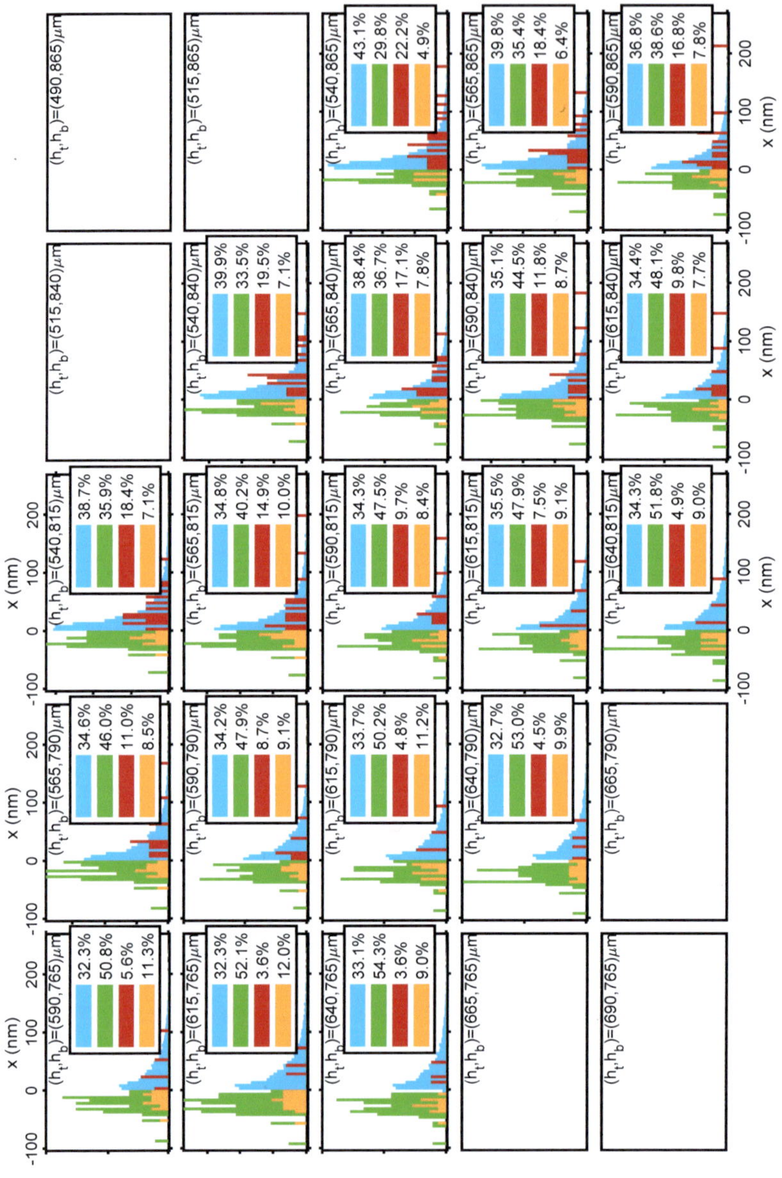

Figure F.7: Defect locations at different interfaces (colors) as deduced from analysis runs for varying distances h_t and h_b between the qubit and the top and bottom electrodes around their nominal values $h_t = 590\,\mu$m and $h_b = 815\,\mu$m, and with a variation step of $25\,\mu$m. The legends indicate the participation of the given interface to the total amount of detected field-tunable TLS. The interface participations and errors quoted in Eq. (4.7) of the main text result from the arithmetic mean and standard deviation of equivalent results obtained with a smaller variation step of $10\,\mu$m. The empty frames denote combinations outside of the maximum possible determination error $\pm 50\,\mu$m to the electrode distances. Further, the distance variations obey the fact that h_t depends on changes in h_b due to the sample holder geometry (see Fig. 3.9 c), but not vice versa. For example, if we choose h_t to be shortened by $50\,\mu$m and h_b by $25\,\mu$m, the resulting distances will be $(h_t - 50\,\mu\text{m} + 25\,\mu\text{m}, h_b - 25\,\mu\text{m}) = (565, 790)\,\mu$m.

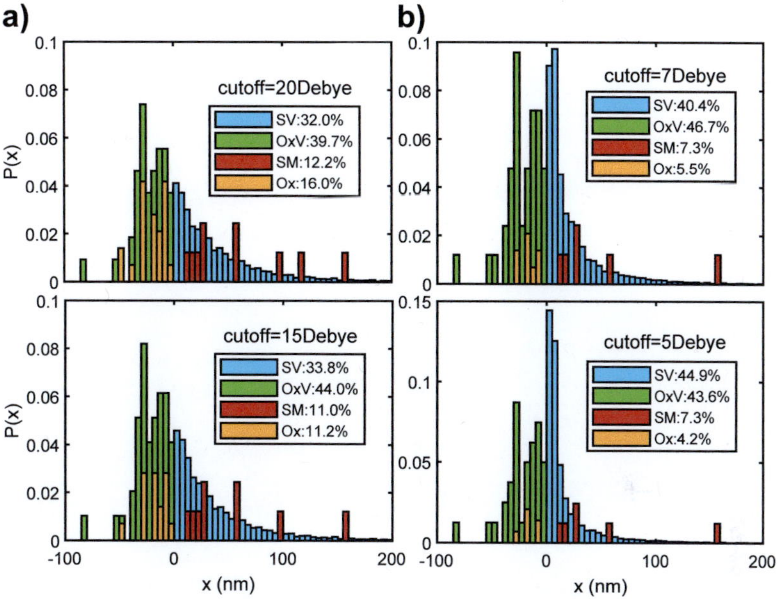

Figure F.8: TLS location histograms and interface contributions (legends) deduced for different cutoffs by which the TLS locations are truncated. **a** Compared to the 10 D cutoff applied in this work (see Fig. 4.14), the increased cutoffs lead to increased participation of the Ox interface, which is expected since most truncated location solutions for the applied 10 D cutoff are at the Ox interface (c.f. Fig. 4.12 d). **b** For cutoffs below 10 D, the histograms become narrower around $x \approx 0$ (substrate-metal-vacuum edge), which is expected due to strongest qubit fields in this region. Further, decreased cutoffs lead to increased participation of the SV-interface. The reason stems from the method to calculate the probability for a single TLS to reside at a given interface, explained in Fig. 2.11: TLS locations at film interfaces (OxV, SM, Ox) are degenerate in the TLS orientation and thus, they are usually stronger weighted than the orientation-specific solutions at SV-interfaces. On the other hand, in contrast to film interfaces, solutions at SV-interfaces exist for every analyzed TLS even for very small cutoffs. Hence, if all film interface solutions of a TLS happen to be discarded due to a small cutoff value, the SV solutions are maximally weighted which leads to the peak at $x \approx 0$ in the SV histograms shown in this figure. To conclude, the 10 D cutoff is physically motivated: it is the maximum imaginable dipole moment size in solids, and there is no reason to use larger cutoffs. However, smaller cutoffs lead to loss of reasonable solutions and to an unrealistic discontinuity of the location histogram at $x \approx 0$.

Bibliography

[Aga+13] K. Agarwal, I. Martin, M. D. Lukin, and E. Demler: *Polaronic model of two-level systems in amorphous solids*. Phys. Rev. B **87** (2013), p. 144201. DOI: 10.1103/PhysRevB.87.144201 (cit. on pp. 1, 10, 11, 49).

[AHV72] P. W. Anderson, B. I. Halperin, and C. Varma: *Anomalous low-temperature thermal properties of glasses and spin glasses*. Philos. Mag. **25** (1972), p. 1 (cit. on pp. 1, 10, 12).

[Bar+13] R. Barends, J. Kelly, A. Megrant, D. Sank, E. Jeffrey, Y. Chen, Y. Yin, B. Chiaro, J. Mutus, C. Neill, P. O'Malley, P. Roushan, J. Wenner, T. C. White, A. N. Cleland, and J. M. Martinis: *Coherent Josephson Qubit Suitable for Scalable Quantum Integrated Circuits*. Phys. Rev. Lett. **111** (2013), p. 080502. DOI: 10.1103/PhysRevLett.111.080502 (cit. on pp. 1, 3, 8, 13, 14, 18, 44, 50–52, 55–57, 61, 66, 68).

[Bie16] A. Bienfait: *Magnetic resonance with quantum microwaves. Quantum Physics. Université Paris-Saclay, 2016. English. NNT: 2016SACLS297. tel-01496176* (2016) (cit. on p. 91).

[Bie+17] A. Bienfait, P. Campagne-Ibarcq, A. H. Kiilerich, X. Zhou, S. Probst, J. J. Pla, T. Schenkel, D. Vion, D. Esteve, J. J. L. Morton, K. Moelmer, and P. Bertet: *Magnetic Resonance with Squeezed Microwaves*. Phys. Rev. X **7** (2017), p. 041011. DOI: 10.1103/PhysRevX.7.041011 (cit. on p. 91).

[Bil19] A. Bilmes: *Additional data for the dissertation: Resolving locations of defects in superconducting transmon qubits* (2019). DOI: 10.5445/IR/1000097356 (cit. on p. 95).

[Bil+17] A. Bilmes, S. Zanker, A. Heimes, M. Marthaler, G. Schön, G. Weiss, A. V. Ustinov, and J. Lisenfeld: *Electronic decoherence of two-level systems in a Josephson junction*. Phys. Rev. B **96** (2017), p. 064504. DOI: 10.1103/PhysRevB.96.064504 (cit. on pp. 12, 43, 46).

[Bla81] J. L. Black: *Low-energy excitations in metallic glasses*. Glassy Metals I, Topics in Applied Physics **46** (1981), pp. 167–190. DOI: 10.1007/3540104402_8 (cit. on pp. 10, 12).

[Bla+04] A. Blais, R.-S. Huang, A. Wallraff, S. M. Girvin, and R. J. Schoelkopf: *Cavity quantum electrodynamics for superconducting electrical circuits: An architecture for quantum computation*. Phys. Rev. A **69** (2004), p. 062320. DOI: 10.1103/PhysRevA.69.062320 (cit. on p. 9).

[Bre+17] J. D. Brehm, A. Bilmes, G. Weiss, A. V. Ustinov, and J. Lisenfeld: *Transmission-line resonators for the study of individual two-level tunneling systems*. Applied Physics Letters **111**.11 (2017), p. 112601. DOI: 10.1063/1.5001920 (cit. on pp. 2, 45, 47, 48).

[Cap+18] T. Capelle, E. Flurin, E. Ivanov, J. Palomo, M. Rosticher, S. Chua, T. Briant, P.-F. Cohadon, A. Heidmann, T. Jacqmin, and S. Deleglise: *Energy relaxation properties of a microwave resonator coupled to a pumped two-level system bath*. arXiv 1805.04397 (2018). URL: https://arxiv.org/abs/1805.04397 (cit. on pp. 1, 12).

[Cat+11] G. Catelani, J. Koch, L. Frunzio, R. J. Schoelkopf, M. H. Devoret, and L. I. Glazman: *Quasiparticle Relaxation of Superconducting Qubits in the Presence of Flux*. Phys. Rev. Lett. **106** (2011), p. 077002. DOI: 10.1103/PhysRevLett.106.077002 (cit. on pp. 1, 16, 68).

[Cho+09] S. Choi, D.-H. Lee, S. G. Louie, and J. Clarke: *Localization of Metal-Induced Gap States at the Metal-Insulator Interface: Origin of Flux Noise in SQUIDs and Superconducting Qubits*. Phys. Rev. Lett. **103** (2009), p. 197001. DOI: 10.1103/PhysRevLett.103.197001 (cit. on pp. 1, 10, 49).

[Chu+16] Y. Chu, C. Axline, C. Wang, T. Brecht, Y. Y. Gao, L. Frunzio, and R. J. Schoelkopf: *Suspending superconducting qubits by silicon micromachining*. Applied Physics Letters **109**.11 (2016), p. 112601. DOI: 10.1063/1.4962327 (cit. on pp. 1, 69, 81).

[Coo+04] K. B. Cooper, M. Steffen, R. McDermott, R. W. Simmonds, S. Oh, D. A. Hite, D. P. Pappas, and J. M. Martinis: *Observation of Quantum Oscillations between a Josephson Phase Qubit and a Microscopic Resonator Using Fast Readout*. Phys. Rev. Lett. **93** (2004), p. 180401. DOI: 10.1103/PhysRevLett.93.180401 (cit. on pp. 2, 13).

[Cór+11] A. D. Córcoles, J. M. Chow, J. M. Gambetta, C. Rigetti, J. R. Rozen, G. A. Keefe, M. Beth Rothwell, M. B. Ketchen, and M. Steffen: *Protecting superconducting qubits from radiation*. Applied Physics Letters **99**.18 (2011), p. 181906. DOI: 10.1063/1.3658630 (cit. on p. 28).

[DDH87] D. A. Danner, M. Dalvie, and D. W. Hess: *Plasma Etching of Aluminum: A Comparison of Chlorinated Etchants*. Journal of The Electrochemical Society **134**.3 (1987), pp. 669–673. DOI: 10.1149/1.2100529 (cit. on p. 32).

[DOL14] C. Deng, M. Otto, and A. Lupascu: *Characterization of low-temperature microwave loss of thin aluminum oxide formed by plasma oxidation*. Applied Physics Letters **104**.4 (2014), p. 043506. DOI: 10.1063/1.4863686 (cit. on p. 45).

[Dol77] G. J. Dolan: *Offset masks for lift-off photoprocessing*. Applied Physics Letters **31**.5 (1977), pp. 337–339. DOI: 10.1063/1.89690 (cit. on pp. 31, 32).

[Dun+17] A. Dunsworth, A. Megrant, C. Quintana, Z. Chen, R. Barends, B. Burkett, B. Foxen, Y. Chen, B. Chiaro, A. Fowler, R. Graff, E. Jeffrey, J. Kelly, E. Lucero, J. Y. Mutus, M. Neeley, C. Neill, P. Roushan, D. Sank, A. Vainsencher, J. Wenner, T. C. White, and J. M. Martinis: *Characterization and reduction of capacitive loss induced by sub-micron Josephson junction fabrication in superconducting qubits.* Applied Physics Letters **111**.2 (2017), p. 022601. DOI: 10.1063/1.4993577 (cit. on pp. 68, 69, 81).

[Dut+12] S. Dutta, J. M. Biser, R. P. Vinci, and H. M. Chan: *Solid State Annealing Behavior of Aluminum Thin Films on Sapphire.* Journal of the American Ceramic Society **95**.2 (2012), pp. 823–830. DOI: 10.1111/j.1551-2916.2011.05011.x (cit. on pp. 73, 74).

[FI06] L. Faoro and L. B. Ioffe: *Quantum Two Level Systems and Kondo-Like Traps as Possible Sources of Decoherence in Superconducting Qubits.* Phys. Rev. Lett. **96** (2006), p. 047001. DOI: 10.1103/PhysRevLett.96.047001 (cit. on pp. 1, 10, 11, 49).

[FI15] L. Faoro and L. B. Ioffe: *Interacting tunneling model for two-level systems in amorphous materials and its predictions for their dephasing and noise in superconducting microresonators.* Phys. Rev. B **91** (2015), p. 014201. DOI: 10.1103/PhysRevB.91.014201 (cit. on p. 2).

[Gam+17] J. M. Gambetta, C. E. Murray, Y. Fung, D. T. McClure, O. Dial, W. Shanks, J. W. Sleight, and M. Steffen: *Investigating Surface Loss Effects in Superconducting Transmon Qubits.* IEEE Transactions on Applied Superconductivity **27**.1 (2017), pp. 1–5. DOI: 10.1109/TASC.2016.2629670 (cit. on p. 1).

[Gor+14] L. Gordon, H. Abu-Farsakh, A. Janotti, and C. G. Van de Walle: *Hydrogen bonds in Al2O3 as dissipative two-level systems in superconducting qubits.* Scientific Reports **4** (2014), pp. 7590–. URL: https://doi.org/10.1038/srep07590 (cit. on pp. 1, 10, 11, 68).

[Gra+17] S. E. de Graaf, A. A. Adamyan, T. Lindström, D. Erts, S. E. Kubatkin, A. Y. Tzalenchuk, and A. V. Danilov: *Direct Identification of Dilute Surface Spins on* Al_2O_3*: Origin of Flux Noise in Quantum Circuits.* Phys. Rev. Lett. **118** (2017), p. 057703. DOI: 10.1103/PhysRevLett.118.057703 (cit. on pp. 1, 2, 10, 11, 68).

[Gra+18] S. E. de Graaf, L. Faoro, J. Burnett, A. A. Adamyan, A. Y. Tzalenchuk, S. E. Kubatkin, T. Lindström, and A. V. Danilov: *Suppression of low-frequency charge noise in superconducting resonators by surface spin desorption.* Nature Communications **9**.1 (2018), pp. 1143–. URL: https://doi.org/10.1038/s41467-018-03577-2 (cit. on p. 2).

[Gra+12] G. J. Grabovskij, T. Peichl, J. Lisenfeld, G. Weiss, and A. V. Ustinov: *Strain Tuning of Individual Atomic Tunneling Systems Detected by a Superconducting Qubit.* Science **338**.6104 (2012), pp. 232–234. DOI: 10.1126/science.1226487 (cit. on pp. 43, 45).

[Gra14] G. J. Grabovskij: *Investigation of coherent microscopic defects inside the tunneling barrier of a Josephson junction*. Karlsrue: KIT Scientific Publishing, 2014, ISBN: 9783731502104 (cit. on pp. 38, 39).

[Gr17] L. Grünhaupt, U. von Lüpke, D. Gusenkova, S. T. Skacel, N. Maleeva, S. Schlör, A. Bilmes, H. Rotzinger, A. V. Ustinov, M. Weides, and I. M. Pop: *An argon ion beam milling process for native AlOx layers enabling coherent superconducting contacts*. Applied Physics Letters **111**.7 (2017), p. 072601. DOI: 10.1063/1.4990491 (cit. on pp. 32, 35).

[Gr18] L. Grünhaupt, M. Spiecker, D. Gusenkova, N. Maleeva, S. T. Skacel, I. Takmakov, F. Valenti, P. Winkel, H. Rotzinger, A. V. Ustinov, and I. M. Pop: *Granular aluminum: A superconducting material for high impedance quantum circuits*. arXiv 1810.12801 (2018). URL: https://arxiv.org/abs/1809.10646 (cit. on p. 2).

[Hes82] D. W. Hess: *Plasma etch chemistry of aluminum and aluminum alloy films*. Plasma Chemistry and Plasma Processing **2**.2 (1982), pp. 141–155. DOI: 10.1007/BF00633130 (cit. on pp. 32, 73, 74).

[Hol+13] A. M. Holder, K. D. Osborn, C. J. Lobb, and C. B. Musgrave: *Bulk and Surface Tunneling Hydrogen Defects in Alumina*. Phys. Rev. Lett. **111** (2013), p. 065901. DOI: 10.1103/PhysRevLett.111.065901 (cit. on pp. 1, 10, 11).

[JC63] E. T. Jaynes and F. W. Cummings: *Comparison of quantum and semiclassical radiation theories with application to the beam maser*. Proceedings of the IEEE **51**.1 (1963), pp. 89–109. DOI: 10.1109/PROC.1963.1664 (cit. on p. 9).

[Kim+08] Z. Kim, V. Zaretskey, Y. Yoon, J. F. Schneiderman, M. D. Shaw, P. M. Echternach, F. C. Wellstood, and B. S. Palmer: *Anomalous avoided level crossings in a Cooper-pair box spectrum*. Phys. Rev. B **78** (2008), p. 144506. DOI: 10.1103/PhysRevB.78.144506 (cit. on p. 1).

[KMM95] A. W. Kleinsasser, R. E. Miller, and W. H. Mallison: *Dependence of critical current density on oxygen exposure in Nb-AlO/sub x/-Nb tunnel junctions*. IEEE Transactions on Applied Superconductivity **5**.1 (1995), pp. 26–30. DOI: 10.1109/77.384565 (cit. on p. 33).

[Kli+18] P. V. Klimov, J. Kelly, Z. Chen, M. Neeley, A. Megrant, B. Burkett, R. Barends, K. Arya, B. Chiaro, Y. Chen, A. Dunsworth, A. Fowler, B. Foxen, C. Gidney, M. Giustina, R. Graff, T. Huang, E. Jeffrey, E. Lucero, J. Y. Mutus, O. Naaman, C. Neill, C. Quintana, P. Roushan, D. Sank, A. Vainsencher, J. Wenner, T. C. White, S. Boixo, R. Babbush, V. N. Smelyanskiy, H. Neven, and J. M. Martinis: *Fluctuations of Energy-Relaxation Times in Superconducting Qubits*. Phys. Rev. Lett. **121** (2018), p. 090502. DOI: 10.1103/PhysRevLett.121.090502 (cit. on pp. 1, 14, 63).

[Koc+07] J. Koch, T. M. Yu, J. Gambetta, A. A. Houck, D. I. Schuster, J. Majer, A. Blais, M. H. Devoret, S. M. Girvin, and R. J. Schoelkopf: *Charge-insensitive qubit design derived from the Cooper pair box*. Phys. Rev. A **76** (2007), p. 042319. DOI: 10.1103/PhysRevA.76.042319 (cit. on pp. 6, 7, 9, 44).

[KVM07] P. J. Koppinen, L. M. Väistö, and I. J. Maasilta: *Complete stabilization and improvement of the characteristics of tunnel junctions by thermal annealing*. Applied Physics Letters **90**.5 (2007), p. 053503. DOI: 10.1063/1.2437662 (cit. on pp. 33, 34).

[Kum+16] P. Kumar, S. Sendelbach, M. A. Beck, J. W. Freeland, Z. Wang, H. Wang, C. C. Yu, R. Q. Wu, D. P. Pappas, and R. McDermott: *Origin and Reduction of* $1/f$ *Magnetic Flux Noise in Superconducting Devices*. Phys. Rev. Applied **6** (2016), p. 041001. DOI: 10.1103/PhysRevApplied.6.041001 (cit. on pp. 1, 2, 10, 11, 68).

[Lec+11] F. Lecocq, I. M. Pop, Z. Peng, I. Matei, T. Crozes, T. Fournier, C. Naud, W. Guichard, and O. Buisson: *Junction fabrication by shadow evaporation without a suspended bridge*. Nanotechnology **22**.31 (2011), p. 315302. URL: http://stacks.iop.org/0957-4484/22/i=31/a=315302 (cit. on pp. 36–38, 74).

[Lis+10] J. Lisenfeld, C. Müller, J. H. Cole, P. Bushev, A. Lukashenko, A. Shnirman, and A. V. Ustinov: *Rabi spectroscopy of a strongly driven qubit-fluctuator system*. Phys. Rev. B **81** ((2010)), 100511(R) (cit. on pp. 2, 13).

[Lis+19] J. Lisenfeld, A. Bilmes, A. Megrant, R. Barends, J. Kelly, P. Klimov, G. Weiss, J. M. Martinis, and A. V. Ustinov: *Electric field spectroscopy of material defects in transmon qubit*. arXiv 1909.09749 (2019). URL: https://arxiv.org/abs/1909.09749 (cit. on pp. 61, 68).

[Lis+15] J. Lisenfeld, G. J. Grabovskij, C. Müller, J. H. Cole, G. Weiss, and A. V. Ustinov: *Observation of directly interacting coherent two-level systems in an amorphous material*. Nat. Commun. **6** (2015), p. 6182. URL: http://dx.doi.org/10.1038/ncomms7182 (cit. on p. 2).

[LU08] A. Lukashenko and A. V. Ustinov: *Improved powder filters for qubit measurements*. Review of Scientific Instruments **79**.1 (2008), p. 014701. DOI: 10.1063/1.2827515 (cit. on p. 29).

[Mar+05] J. M. Martinis, K. B. Cooper, R. McDermott, M. Steffen, M. Ansmann, K. D. Osborn, K. Cicak, S. Oh, D. P. Pappas, R. W. Simmonds, and C. C. Yu: *Decoherence in Josephson Qubits from Dielectric Loss*. Phys. Rev. Lett. **95** (2005), p. 210503. DOI: 10.1103/PhysRevLett.95.210503 (cit. on pp. 1, 10–12, 43, 45, 46, 49, 68, 88).

[MM14] J. M. Martinis and A. Megrant: *UCSB final report for the CSQ program: Review of decoherence and materials physics for superconducting qubits* (2014), arXiv:1410.5793 (cit. on p. 81).

[Mat+19] S. Matityahu, H. Schmidt, A. Bilmes, A. Shnirman, G. Weiß, A. V. Ustinov, M. Schechter, and J. Lisenfeld: *Energy relaxation properties of a microwave resonator coupled to a pumped two-level system bath*. arXiv 1903.07914 (2019). URL: https://arxiv.org/abs/1903.07914 (cit. on p. 2).

[Mat+16] S. Matityahu, A. Shnirman, G. Schön, and M. Schechter: *Decoherence of a quantum two-level system by spectral diffusion*. Phys. Rev. B **93** (2016), p. 134208. DOI: 10.1103/PhysRevB.93.134208 (cit. on p. 2).

[Mei+18] S. M. Meißner, A. Seiler, J. Lisenfeld, A. V. Ustinov, and G. Weiss: *Probing individual tunneling fluctuators with coherently controlled tunneling systems*. Phys. Rev. B **97** (2018), p. 180505. DOI: 10.1103/PhysRevB.97.180505 (cit. on pp. 2, 63).

[MCJ17] C. Müller, J. H. Cole, and L. J.: *Towards understanding two-level-systems in amorphous solids - Insights from quantum circuits* (2017), arXiv:1705.01108v2 (cit. on p. 11).

[Mül+14] C. Müller, J. Lisenfeld, A. Shnirman, and S. Poletto: "Interacting two-level defects as sources of fluctuating high-frequency noise in superconducting circuits". in preparation. 2014 (cit. on pp. 1, 14, 63).

[NH07] J. R. Nesbitt and A. F. Hebard: *Time-dependent glassy behavior of interface states in* $Al-AlO_x-Al$ *tunnel junctions*. Phys. Rev. B **75** (2007), p. 195441. DOI: 10.1103/PhysRevB.75.195441 (cit. on pp. 33, 34).

[NK76] J. Niemeyer and V. Kose: *Observation of large dc supercurrents at nonzero voltages in Josephson tunnel junctions*. Applied Physics Letters **29**.6 (1976), pp. 380–382. DOI: 10.1063/1.89094 (cit. on pp. 31, 32).

[OCo+08] A. D. OConnell, M. Ansmann, R. C. Bialczak, M. Hofheinz, N. Katz, E. Lucero, C. McKenney, M. Neeley, H. Wang, E. M. Weig, A. N. Cleland, and J. M. Martinis: *Microwave dielectric loss at single photon energies and millikelvin temperatures*. Applied Physics Letters **92**.11 (2008), p. 112903. DOI: 10.1063/1.2898887 (cit. on p. 1).

[Pie] *P-882.11 PICMA ® Piezoaktor,* $3\,mm \times 2\,mm \times 9\,mm$*, PI Ceramic GmbH, Lindenstraße, 07589 Lederhose, Germany* (cit. on p. 38).

[Pap+11] D. P. Pappas, M. R. Vissers, D. S. Wisbey, J. S. Kline, and J. Gao: *Two Level System Loss in Superconducting Microwave Resonators*. IEEE Transactions on Applied Superconductivity **21**.3 (2011), pp. 871–874. DOI: 10.1109/TASC.2010.2097578 (cit. on p. 45).

[Pei12] T. Peichl: *Einfluss mechanischer Deformation auf atomare Tunnelsysteme - untersucht mit Josephson Phasen-Qubits*. German. PhD thesis. 2012. 84 pp. DOI: 10.5445/KSP/1000027408 (cit. on pp. 38, 39).

[Phi87] W. A. Phillips: *Two-level states in glasses*. Reports on Progress in Physics **50**.12 (1987), p. 1657. URL: http://stacks.iop.org/0034-4885/50/i=12/a=003 (cit. on p. 11).

[Phi72] W. A. Phillips: *Tunneling states in amorphous solids*. J. Low Temp. Phys. **7**.3-4 (1972), p. 351 (cit. on pp. 1, 10, 12).

[Pop+12] I. M. Pop, T. Fournier, T. Crozes, F. Lecocq, I. Matei, B. Pannetier, O. Buisson, and W. Guichard: *Fabrication of stable and reproducible submicron tunnel junctions*. Journal of Vacuum Science & Technology B **30**.1 (2012), p. 010607. DOI: 10.1116/1.3673790 (cit. on pp. 33, 34).

[Pop+14] I. M. Pop, K. Geerlings, G. Catelani, R. J. Schoelkopf, L. I. Glazman, and M. H. Devoret: *Coherent suppression of electromagnetic dissipation due to superconducting quasiparticles*. Nature **508** (2014), pp. 369–. URL: https://doi.org/10.1038/nature13017 (cit. on p. 1).

[Pro+99] J. Proost, H. Li, T. Conard, W. Boullart, and K. Maex: *Chemical and Electrical Characterization of the Interaction of BCl3/Cl2 Etching and CF4/H2O Stripping Plasmas with Aluminum Surfaces*. Journal of The Electrochemical Society **146**.11 (1999), pp. 4230–4235. DOI: 10.1149/1.1392619 (cit. on p. 32).

[RL95] I. W. Rangelow and H. Löschner: *Reactive ion etching for microelectrical mechanical system fabrication*. Journal of Vacuum Science & Technology B: Microelectronics and Nanometer Structures Processing, Measurement, and Phenomena **13**.6 (1995), pp. 2394–2399. DOI: 10.1116/1.588007 (cit. on p. 32).

[Ree+10] M. D. Reed, L. DiCarlo, B. R. Johnson, L. Sun, D. I. Schuster, L. Frunzio, and R. J. Schoelkopf: *High-Fidelity Readout in Circuit Quantum Electrodynamics Using the Jaynes-Cummings Nonlinearity*. Phys. Rev. Lett. **105** (2010), p. 173601. DOI: 10.1103/PhysRevLett.105.173601 (cit. on pp. 9, 44, 56).

[Ric+16] C. J. K. Richardson, N. P. Siwak, J Hackley, Z. K. Keane, J. E. Robinson, B Arey, I Arslan, and B. S. Palmer: *Fabrication artifacts and parallel loss channels in metamorphic epitaxial aluminum superconducting resonators*. Superconductor Science and Technology **29**.6 (2016), p. 064003. URL: http://stacks.iop.org/0953-2048/29/i=6/a=064003 (cit. on pp. 32, 73, 74).

[Sar+16] B. Sarabi, A. N. Ramanayaka, A. L. Burin, F. C. Wellstood, and K. D. Osborn: *Projected Dipole Moments of Individual Two-Level Defects Extracted Using Circuit Quantum Electrodynamics*. Phys. Rev. Lett. **116** (2016), p. 167002. DOI: 10.1103/PhysRevLett.116.167002 (cit. on pp. 2, 18, 47, 48).

[Sch+19] S. Schlör, J. Lisenfeld, C. Müller, A. Schneider, D. P. Pappas, A. V. Ustinov, and M. Weides: *Correlating decoherence in transmon qubits: Low frequency noise by single fluctuators*. arXiv 1901.05352 (2019). URL: https://arxiv.org/abs/1901.05352 (cit. on pp. 14, 63).

[Sch+08] J. A. Schreier, A. A. Houck, J. Koch, D. I. Schuster, B. R. Johnson, J. M. Chow, J. M. Gambetta, J. Majer, L. Frunzio, M. H. Devoret, S. M. Girvin, and R. J. Schoelkopf: *Suppressing charge noise decoherence in superconducting charge qubits*. Phys. Rev. B **77** (2008), p. 180502. DOI: 10.1103/PhysRevB.77.180502 (cit. on p. 6).

[Sha+10] Y. Shalibo, Y. Rofe, D. Shwa, F. Zeides, M. Neeley, J. M. Martinis, and N. Katz: *Lifetime and Coherence of Two-Level Defects in a Josephson Junction*. Phys. Rev. Lett. **105** (2010), p. 177001. DOI: 10.1103/PhysRevLett.105.177001 (cit. on pp. 1, 10, 11, 68).

[Sim+04] R. W. Simmonds, K. M. Lang, D. A. Hite, D. P. Pappas, and J. M. Martinis: *Decoherence in Josephson qubits from junction resonances*. Phys. Rev. Lett. **93**.7 (2004), p. 077003 (cit. on p. 1).

[Son+09] C. Song, M. P. DeFeo, K. Yu, and B. L. T. Plourde: *Reducing microwave loss in superconducting resonators due to trapped vortices*. Applied Physics Letters **95**.23 (2009), p. 232501. DOI: 10.1063/1.3271523 (cit. on pp. 16, 68).

[Ste+06] M. Steffen, M. Ansmann, R. McDermott, N. Katz, R. C. Bialczak, E. Lucero, M. Neeley, E. M. Weig, A. N. Cleland, and J. M. Martinis: *State Tomography of Capacitively Shunted Phase Qubits with High Fidelity*. Phys. Rev. Lett. **97** (2006), p. 050502. DOI: 10.1103/PhysRevLett.97.050502 (cit. on p. 1).

[Sue+18] H. le Sueur, A. Svilans, N. Bourlet, A. Murani, L. Bergé, L. Dumoulin, and P. Joyez: *Microscopic charged fluctuators as a limit to the coherence of disordered superconductor devices*. arXiv 1810.12801 (2018). URL: https://arxiv.org/abs/1810.12801 (cit. on p. 2).

[Wan+15] C. Wang, C. Axline, Y. Y. Gao, T. Brecht, Y. Chu, L. Frunzio, M. H. Devoret, and R. J. Schoelkopf: *Surface participation and dielectric loss in superconducting qubits*. Applied Physics Letters **107**.16 (2015), p. 162601. DOI: 10.1063/1.4934486 (cit. on pp. 1, 16, 68).

[Wen+11] J Wenner, M Neeley, R. C. Bialczak, M Lenander, E. Lucero, A. D. O'Connell, D Sank, H Wang, M Weides, A. N. Cleland, and J. M. Martinis: *Wirebond crosstalk and cavity modes in large chip mounts for superconducting qubits*. Superconductor Science and Technology **24**.6 (2011), p. 065001. URL: http://stacks.iop.org/0953-2048/24/i=6/a=065001 (cit. on pp. 41, 42).

[Wu+17] X. Wu, J. L. Long, H. S. Ku, R. E. Lake, M. Bal, and D. P. Pappas: *Overlap junctions for high coherence superconducting qubits*. Applied Physics Letters **111**.3 (2017), p. 032602. DOI: 10.1063/1.4993937 (cit. on p. 81).

List of Publications

Published/ in peer-review process:

1. **A. Bilmes**, A. Megrant, P. Klimov G. Weiss, J. M. Martinis, A. V. Ustinov, and J. Lisenfeld, "Resolving the positions of defects in superconducting quantum bits", arxXiv:1911.08246 (2019).
2. J. Lisenfeld, **A. Bilmes**, A. Megrant, R. Barends, J. Kelly, P. Klimov, G. Weiss, J. M. Martinis, and A. V. Ustinov, "Electric field spectroscopy of material defects in transmon qubit", arXiv:1909.09749 (2019).
3. S. Matityahu, H. Schmidt, **A. Bilmes**, A. Shnirman, G. Weiß, A. V. Ustinov, M. Schechter, and J. Lisenfeld, "Dynamical Decoupling of Quantum Two-Level Systems by Coherent Multiple Landau-Zener Transitions", arXiv:1903.07914 (2019).
4. S. Schlör, J. Lisenfeld, C. Müller, **A. Bilmes**, A. Schneider, D. P. Pappas, A. V. Ustinov, and M. Weides, "Correlating decoherence in transmon qubits: Low frequency noise by single fluctuators", arXiv:1901.07914 (2019).
5. F. Valenti, F. Henriques, G. Catelani, N. Maleeva, L. Grünhaupt, U. von Lüpke, S. T. Skacel, P. Winkel, **A. Bilmes**, A. V. Ustinov, J. Goupy, M. Calvo, A. Benoît, F. Lévy-Bertrand, A. Monfardini, and I. M. Pop, "Interplay between kinetic inductance, non-linearity and quasiparticle dynamics in granular aluminum MKIDs", Phys. Rev. Applied. **11**, 054087 (2019)
6. J.D. Brehm, **A. Bilmes**, G. Weiss, A.V. Ustinov, and J. Lisenfeld, "Transmission line-resonators for the study of individual two-level tunneling systems", Appl. Phys. Lett. **111**, 112601 (2017).
7. **A. Bilmes**, S. Zanker, A. Heimes, M. Marthaler, G. Schön, G. Weiss, A.V. Ustinov, and J. Lisenfeld, *Electronic decoherence of two-level systems in a Josephson junction*, Phys. Rev. B **96**, 064504 (2017)
8. S. Matityahu, J. Lisenfeld, **A. Bilmes**, A. Shnirman, G. Weiss, and A.V. Ustinov, *Rabi noise spectroscopy of individual two-level tunneling defects*, Phys. Rev. B **95**, 241409 (2017)
9. L. Grünhaupt, U. von Lüpke, D. Gusenkova, S. T. Skacel, N. Maleeva, S. Schlör, **A. Bilmes**, H. Rotzinger, A. V. Ustinov, M. Weides, and I. M. Pop, *An argon ion beam milling process for native AlOx layers enabling coherent superconducting contacts*, Appl. Phys. Lett. **111**, 7 (2017)

10. J. Lisenfeld, **A. Bilmes**, S. Matityahu, S. Zanker, M. Marthaler, M. Schechter, G. Schön, A. Shnirman, G. Weiss, and A. V. Ustinov, *Decoherence spectroscopy with individual two-level tunneling defects*, Scientific Reports **7**, 23786 (2016)

In preparation:

1. **A. Bilmes**, J. Brehm, I. Pop, M. Weides, A. V. Ustinov, and J. Lisenfeld, "A Two-Level-System sensor derived from a superconducting qubit", *in preparation* (2019).

Acknowledgments

• Ich danke vielmals Prof. Alexey V. Ustinov für die einzigartige Gelegenheit, meine Doktorarbeit in seiner Gruppe am KIT anzufangen, für die Erstbetreuung meiner Arbeit und die viele Unterstützung. • Ich danke vom ganzen Herzen Dr. Jürgen Lisenfeld für die gemeinsame Durchführung der Projekte dieser Doktorarbeit, die ohne ihn nicht geglückt wäre. Er unterstützte mich unermüdlich in allen technischen und physikalischen Angelegenheiten und führte praktisch alle dieser Arbeit zugrundeliegenden Messungen durch. Des weiteren danke ich ihm für das Lesen all meiner Abstracts und Anträge, vor allem dieser Dissertation. • Ich danke Prof. Alexander Shnirman dafür, mein Koreferent zu sein und für die cleveren Ideen bei Diskussionen. • Ich danke Prof. John Martinis für seine Begeisterung für unsere Experimente und seiner Gruppe für die Qubit Proben, auf denen ein Großteil dieser Arbeit fußt. Des Weiteren danke ich Google Inc. für die finanzielle Unterstützung unserer Forschung am KIT und hoffe auf weitere gemeinsame Projekte. • Ich möchte mich bei Prof. Georg Weiß für seine Ideen bedanken, die grundlegend für unsere vergangenenen und jetzigen Experimente sind. Danke für die vielen inspirierenden Diskussionen und für das sehr detailgetreue Durchlesen meiner Texte. Ich danke ihm, sowie Dr. Torben Peichl, Dr. Grigory Grabovskij und nochmals Dr. Jürgen Lisenfeld für die Entwicklung der noch heute weltweit einzigen Anlage zum elastischen Stimmen von Materialdefekten in Qubits. • Ich danke vielmals Dr. Silvia Diewald und Dipl. Phys. Patrice Brenner für fokussierten Einsatz des Elektronenstahls zwecks Lithographie oder Mikroskopie. Auch möchte ich mich bei Lukas Radtke bedanken, für seinen unermüdlichen Einsatz im Reinraum und super Kollegialität. Ich bedanke mich bei Dr. Aina Quintilla und Heinrich Reimer für ihre Hilfsbereitschaft im Reinraum. • Des weiteren, danke ich Dr. Saskia Meißner und Dr. Arnold Seiler für zahlreiche und detaillierte Diskussionen und Arnold für das Durchlesen meiner Dissertation. Ich hoffe, das künftige gemeinsame Projekt wird uns weiterbringen. • Ich bedanke mich bei M. Sc. Jan Brehm für die ertragreiche Kollaboration während dessen Masterarbeit und für die tollen Ergebnisse. Weiterhin danke ich ihm für inspirierende Diskussionen und für das Lesen meiner Diss. • Ich danke Hans-Willi Pensl sehr für die Versorgung mit Flüssighelium und gute Ratschläge sowie angenehme Unterhaltungen in der Verflüssigungshalle. • Ich danke Michael Meyer und seiner Truppe in der Feinmechanik Werkstatt für die weltklasse-angefertigten Metallteile, und Roland Jehle für perfekt impedanzangepassten Leiterplatinen. • Ich danke Dr. Natasha Maleeva für die Weiterempfehlung von der App Trello, die mich bestens durch das Dissertation-Schreiben gebracht hat. Auch danke ich ihr für die konstruktiven Diskussionen und die sehr heitere Arbeitsatmospäre. • Ich danke Dr. Sebastian Skacel für seine Unterstützung und Ratschläge während der Entstehung dieses Manuskripts. • Ich danke Dr. Hannes Rotzinger für seine vielen guten Ratschläge und wichtige Diskussionen sowie für seine stets kritische Meinung. • Ich danke Dr. Ioan Pop für viele Ratschläge und Tipps bei Fabrikationsproblemen. • Ich danke Steffi Baatz, Birgit Schelske, Ilona Brosch and Claudia Alaya für den herzlichen

Umgang und schnellen administrativen Support. • Ich danke Dr. Oleksandr Lukashenko für seine zahlreichen, pragmatischen und trickreichen Tipps und Support mit der Kryotechnik. • Ich danke Dr. Arkady Fedorov für seine Gastfreundschaft während meines Aufenthalts in seiner Gruppe an der Universität in Brisbane (Au), und ich bedanke mich bei Karlsruhe House of Young Scientists (KHYS) für die Finanzierung dieser Kollaborationsreise. • Ich danke allen jetzigen und ehemaligen Gruppenmitgliedern, vor allem Jan Brehm, Sebastian Skacel und Patrick Winkel für die schönen gemeinsamen Erlebnisse unter anderem während der Tieftemperatur-, Oberflächenwellen- und weiteren wichtigen Seminaren und Konferenzen. • Ich danke Alexander Neumann für das Erklären des Higgs Mechanismus am Vortag meiner Prüfung. • Des weiteren, danke ich allen Mitgliedern des Physikalischen Intituts für die angenehme Arbeits- und Feieratmosphäre. • Bester Dank gilt an die Helmholtz International Research School for Teratronics (KIT) und die Landesgraduiertenförderung (KHYS, KIT) für die Finanzierung dieser Doktorarbeit.

• In Gedenken an Prof. Frank Hekking, der den jungen Praktikanten für Physik von supraleitenden Qubits mit seiner trockenen und unglaublich sympathischen Art inspirieren konnte. • Ich danke vom ganzen Herzen meinem Lehrer Hans Hellmann Junior für dessen Begeisterung für Muster und Symmetrien im Alltag. • In Gedenken an meine Großtante Dilara Alievna Mineeva, die mich von klein auf spielerisch mit elementaren physikalischen Experimenten vertraut machte. • Ich danke dem größten Supporter Florian Keller, dem besten Abitur-Abschluss Redner Volker Haaß und dem coolsten Typen Adrian Kohl für eure Unterstützung bei jeder Wetterlage. • Ich danke meiner fürsorglichen Mutter Annel Bilmes und meinem Vater Vladimir Bilmes, der mir so viel Grundwissen im Kindesalter vermittelte. Für euer Glauben an mich und bedingungslose Liebe. Ohne euch wäre ich nicht dort, wo ich nun stehe. Ich liebe euch. • Tiefster Dank geht an meine geliebte Anna Sophie Rösch und den kleinen Hüpfer Lotta Rösch. Für alle Sicherheit die du mir gabst Anna. Ich liebe dich.

Experimental Condensed Matter Physics
(ISSN 2191-9925)

Herausgeber
Physikalisches Institut

Prof. Dr. David Hunger
Prof. Dr. Alexey Ustinov
Prof. Dr. Georg Weiß
Prof. Dr. Wolfgang Wernsdorfer
Prof. Dr. Wulf Wulfhekel

Band 1 Alexey Feofanov
Experiments on flux qubits with pi-shifters. 2011
ISBN 978-3-86644-644-1

Band 2 Stefan Schmaus
Spintronics with individual metal-organic molecules. 2011
ISBN 978-3-86644-649-6

Band 3 Marc Müller
Elektrischer Leitwert von magnetostriktiven Dy-Nanokontakten. 2011
ISBN 978-3-86644-726-4

Band 4 Torben Peichl
Einfluss mechanischer Deformation auf atomare Tunnelsysteme – untersucht mit Josephson Phasen-Qubits. 2012
ISBN 978-3-86644-837-7

Band 5 Dominik Stöffler
Herstellung dünner metallischer Brücken durch Elektromigration und Charakterisierung mit Rastersondentechniken. 2012
ISBN 978-3-86644-843-8

Band 6 Tihomir Tomanic
Untersuchung des elektronischen Oberflächenzustands von Ag-Inseln auf supraleitendem Niob (110). 2012
ISBN 978-3-86644-898-8

Band 7 Lukas Gerhard
Magnetoelectric coupling at metal surfaces. 2013
ISBN 978-3-7315-0063-6

Band 8 Nicht erschienen.

Die Bände sind unter www.ksp.kit.edu als PDF frei verfügbar oder als Druckausgabe bestellbar.

Band 9	Jochen Zimmer **Cooper pair transport in arrays of Josephson junctions.** 2014 ISBN 978-3-7315-0130-5
Band 10	Oliver Berg **Elektrischer Transport durch Nanokontakte von Selten-Erd-Metallen.** 2014 ISBN 978-3-7315-0209-8
Band 11	Grigorij Jur'evic Grabovskij **Investigation of coherent microscopic defects inside the tunneling barrier of a Josephson junction.** 2014 ISBN 978-3-7315-0210-4
Band 12	Cornelius Thiele **STM Characterization of Phenylene-Ethynylene Oligomers on Au(111) and their Integration into Carbon Nanotube Nanogaps.** 2014 ISBN 978-3-7315-0235-7
Band 13	Michael Peter Schackert **Scanning Tunneling Spectroscopy on Electron-Boson Interactions in Superconductors.** 2014 ISBN 978-3-7315-0238-8
Band 14	Susanne Butz **One-Dimensional Tunable Josephson Metamaterials.** 2014 ISBN 978-3-7315-0271-5
Band 15	Philipp Jung **Nonlinear Effects in Superconducting Quantum Interference Meta-Atoms.** 2014 ISBN 978-3-7315-0294-4
Band 16	Sebastian Probst **Hybrid quantum system based on rare earth doped crystals.** 2016 ISBN 978-3-7315-0345-3
Band 17	Wolfram Kittler **Magnetische Anisotropie und Quantenphasenübergang in $CeTi_{1-x}V_xGe_3$.** 2015 ISBN 978-3-7315-0363-7
Band 18	Moritz Peter **Towards magnetic resonance in scanning tunneling microscopy using heterodyne detection.** 2015 ISBN 978-3-7315-0410-8

Band 19 Junji Tobias Märkl
Investigation of Magnetic Adatoms with Scanning Tunneling Techniques. 2015
ISBN 978-3-7315-0435-1

Band 20 Jochen Braumüller
Quantum simulation experiments with superconducting circuits. 2018
ISBN 978-3-7315-0780-2

Band 21 Jinjie Chen
Local Investigation of Single Magnetic Molecules with Scanning Tunneling Microscopy. 2018
ISBN 978-3-7315-0819-9

Band 22 Arnold Seiler
Einfluss der Leitungselektronen auf die Dynamik atomarer Tunnelsysteme in ungeordneten Festkörpern: Relaxationsprozesse in metallischen Gläsern und ungeordneten dünnen Aluminiumoxid-Schichten. 2019
ISBN 978-3-7315-0870-0

Band 23 Jasmin Maria Jandke
Elastic and Inelastic Scanning Tunneling Spectroscopy on Iron-Based Superconductors. 2019
ISBN 978-3-7315-0747-5

Band 24 Kevin Edelmann
Electroluminescence from Plasmonic Excitations in a Scanning Tunnelling Microscope. 2019
ISBN 978-3-7315-0923-3

Band 25 Lei Zhang
Sub-Kelvin scanning tunneling microscopy on magnetic molecules. 2019
ISBN 978-3-86644-950-3

Band 26 Lukas Grünhaupt
Granular aluminium superinductors. 2019
ISBN 978-3-7315-0956-1

Band 27 Alexander Bilmes
Resolving locations of defects in superconducting transmon qubits. 2019
ISBN 978-3-7315-0967-7